Photography, Temporality, and Modernity

This book examines photography's unique capacity to represent time with a degree of elasticity and abstraction. Part object-study, part cultural/philosophical history, it examines the medium's ability to capture and sometimes "defy" time, while also traveling as objects across time-and-space nexuses. The book features studies of understudied, widespread, practices: studio portraiture, motion studies, panoramas, racing photo finishes, composite college class pictures, planetary photography, digital montages, and extended-exposure images. A closer look at these images and their unique cultural/historical contexts reveals photography to be a unique medium for expressing changing perceptions of time, and the anxiety its passage provokes.

Kris Belden-Adams is an Assistant Professor of Art History at the University of Mississippi, and specializes in the history and theory of photography and contemporary art. She is the author of *Photography, Eugenics, "Aristogenics": Picturing Privilege* (2019), and the editor and contributor of two chapters to the volume *Photography and Failure: One Medium's Entanglement With Mishaps, Flops, and Disappointments* (2017).

Routledge History of Photography

This series publishes research monographs and edited collections focusing on the history and theory of photography. These original, scholarly books may take an art historical, visual studies, or material studies approach.

For a full list of titles in this series, please visit https://www.routledge.com/Routledge-History-of-Photography/book-series/RHOP

1. **Liminalities of Gender and Sexuality in Nineteenth-Century Iranian Photography**
   Desirous Bodies
   Staci Gem Scheiwiller

2. **Travel Marketing and Popular Photography in Britain, 1888–1939**
   Reading the Travel
   Sara Dominici

3. **The Pioneering Photographic Work of Hercule Florence**
   Boris Kossoy

4. **Photography and Ontology**
   Unsettling Images
   Edited by Donna West Brett and Natalya Lusty

Photography, Temporality, and Modernity

Time Warped

Kris Belden-Adams

Routledge
Taylor & Francis Group
LONDON AND NEW YORK

First published 2019 by Routledge

2 Park Square, Milton Park, Abingdon, Oxon OX14 4RN
605 Third Avenue, New York, NY 10017

Routledge is an imprint of the Taylor & Francis Group, an informa business

First issued in paperback 2022

Publisher's Note

The publisher has gone to great lengths to ensure the quality of this reprint but points out that some imperfections in the original copies may be apparent.

Library of Congress Cataloging-in-Publication Data
Names: Belden-Adams, Kris, author.
Title: Photography, temporality, and modernity : time warped / Kris Belden-Adams.
Description: First edition. | New York, NY : Routledge, 2019. | Series: Routledge history of photography | Includes bibliographical references and index.
Identifiers: LCCN 2018044798| ISBN 9781138544314 (hardback : alk. paper) | ISBN 9781138544321 (ebook)
Subjects: LCSH: Time and photography. | Photography–Psychological aspects. | Time–Psychological aspects.
Classification: LCC TR183 .B424 2019 | DDC 770–dc23
LC record available at https://lccn.loc.gov/2018044798

ISBN: 978-1-138-54431-4 (hbk)
ISBN: 978-1-03-233861-3 (pbk)
DOI: 10.4324/9781138544321

Typeset in Sabon
by Swales & Willis Ltd

This book is dedicated to Lily and Avery, whose experiences with time will surpass mine.

Contents

Figures

Acknowledgements

Any project of this scope is necessarily a collaborative effort. I would like to offer thanks to several individuals and institutions who have helped enable my research and writing.

First, a University of Mississippi College of Liberal Arts Summer Research Grant made it possible for me to illustrate this book with photographs. It would be a lot less interesting, visually, without the college's support.

Special thanks go to William Stefanov of the National Aeronautics and Space Administration for demystifying the process by which whole-Earth digital images were made, and for helping me understand the eventual transition of this image-making practice to private for-profit companies such as Google/Planet.

I also would like to thank Andrew Davidhazy for helping me understand how a racing photo finish is made. I cannot overstate my gratitude for his warmth and generosity in sharing his research and photographs with me. If all academicians believed as he does – that knowledge is to be shared, not coveted – the scholarly world would be the richer for it.

On that note, several colleagues in my field have helped me hone my ideas as this project has evolved. Among them, I thank Geoffrey Batchen and Allison Moore for helping to sharpen my thinking. I thank University of Mississippi colleague Tyler Barnes for contributing informational graphics to help illustrate this book.

And finally, I thank John for supporting this project from its genesis.

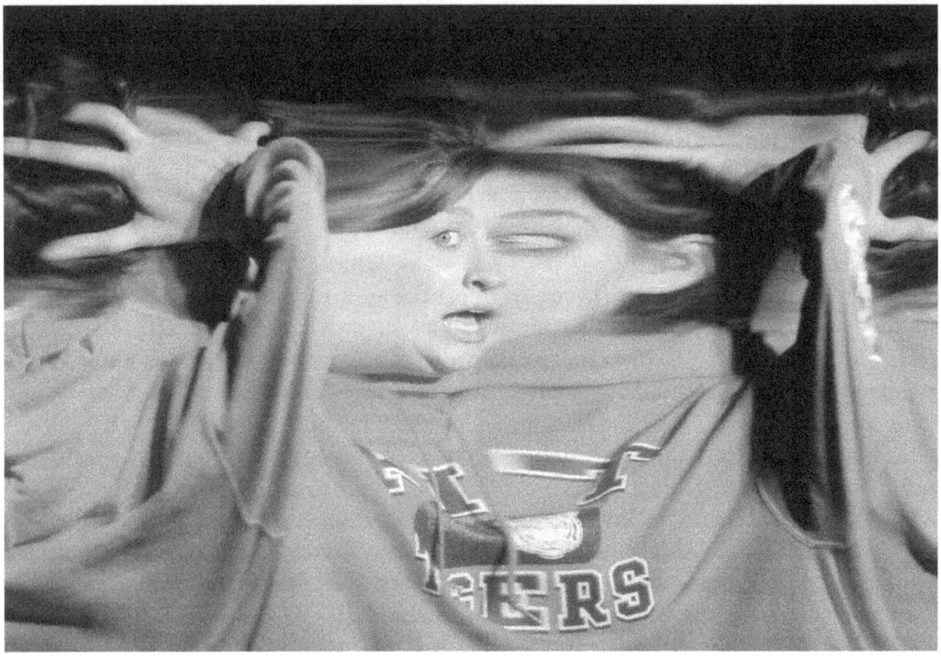

Figure 0.1 Cover: Andrew Davidhazy, *Carlissa*, 2009. © Andrew Davidhazy

1 Photography and Time

Many theorists, historians, and commentators have noted that photography has a distinctive and complex manner of representing temporality. Roland Barthes, for example, wrote that the photograph has a peculiar capacity to represent the past in the present, and thus to imply the passing of time in general.[1] As a consequence, Barthes argued, all photographs speak of the inevitability of our own death in the future. Moreover, he linked photography's peculiar, tense-confounding temporality to its capacity for a certain kind of realism: "false on the level of perception, true on the level of time."[2] Barthes's analysis poses a challenge to later commentators on photography to explore these questions about the medium: What exactly is photography's complex, multifarious relationship to time? How might it express and reflect the human condition in the ages of technological modernity and postmodernity?

This book addresses those queries by analyzing in detail a sample of understudied, mass-accessible, commonplace vernacular photographic practices, positioned within the temporal mindsets of their respective cultures, and considered as expressions of their own shifting contexts. Rather than provide a comprehensive, and necessarily incomplete, study of every possible way in which photography may express time, this study instead focuses on in-depth analyses of specific photographic practices, including stroboscopic images, fin-de-siècle family studio portraiture, racing photo finishes, and the National Aeronautics and Space Administration's (NASA's) whole-Earth, digital-composite images. It will be argued that each of these photographic practices represents time, as it is filtered through their respective, unique temporal cultures, in at least three distinct ways: as atomized time, narrative time, elapsed/"sculptural" time, and as synthetic/hypothetical time. Each chapter offers localized historical views of those photographic practices and their contexts, in an effort to explore the evolving cultures of temporality that gave form to those photographic practices.

Specifically, Chapter 2 will begin by taking a closer look at the temporal climate into which photography was born. It will examine early written accounts and photographs by several pioneers of the medium, including William Henry Fox Talbot, and Louis Jacques Mandé Daguerre, whose work – along with then-popular Romantic literature and painting, and the development of photographic technology – helps to establish an "atomized" view of a fraction-of-a-second, full-field of space as *the way* by which photography strove to convey time. This approach consequently also became a dominant narrative thread in our histories of the medium. The founders and the earliest commentators on photography noted the medium's peculiar capacity to capture a thin slice of time, and they noted that photography's unique views differed dramatically from the appearance of known reality, in which time's relentless passage is non-negotiable. For the "lived" experience of time does not halt, slow, or enable itself to be reversed, revisited, or re-lived for the sake of playing through different consequences.

Yet a closer look at Harold Edgerton's stroboscopic work produced at the Massachusetts Institute of Technology – which represents an apex achievement of the "atomized" visual slice of photographic time by capturing 1,000,000th-of-a-second durations – reveals such views to be far more temporally complex and multifarious than they may at first appear. When Edgerton's *Death of a Light Bulb* series is examined in relationship to Barthes's systems of "relay" and *diegesis* to describe the production of narratives in visual culture, even the most "instantaneous" of "atomized" photographic views, born at the hands of an electrical engineer at MIT, is riddled with ambiguities.

To further study the role of narrativity in temporalizing photographs that is introduced in Chapter 2, the chapter that follows will examine the complex experiences of time offered by a fin-de-siècle, studio-made, family portrait from the Oklahoma territory. This commonplace, straightforward, vernacular photographic practice is often overlooked in our histories, despite its ubiquity and complexity as a visual form. In Chapter 3, a specific family portrait, the *Leighton Family* (1899), is compared to a poem about the making of the photograph written many years later by one of its subjects, Nancy Ellen Miller, who was just six years old at the time the portrait was made. The image's connection to broader historical narratives, such as the settling of the Oklahoma frontier and the Dalton Gang of outlaws (the Leightons' frontier neighbors), is explored and positioned in relation to unpublished poems by Miller. In addition, the *Leighton Family* is contextualized with respect to the unique challenges posed to the cohesivity of the family unit by the mobility encouraged by the first railroad lines – which themselves reconfigured conceptions of space and time.

When compared to a Xeroxed, "updated" version of the *Leighton Family* portrait, made by Billie Hazel Leighton Flora to include a sibling who had not yet been born (her father, William Orman Leighton), additional, fluid temporal associations emerge. The "narrative times" evoked by portraits such as the *Leighton Family* operate in the realm of the subjective, in order to aid in the building and maintenance of personal memories and histories. Such stories, like the human experience itself, are amalgamations comprised of conflicting temporalities, built upon the surprisingly steady, yet essentially flawed, faculty of memory, tinged with optimism, nostalgia, and hope for the future. The *Leighton Family*, it is argued, offers viewers non-linear, multiple temporal vacillations between past, present, and future tenses (and combinations of these). This exemplary portrait has a complicated, unfixed, and potentially endless number of ways to relate to temporality that far exceed Barthes's conception of future-anterior time.

Chapter 4 explores another possible relationship of photography to temporality, through a case study of the unique manipulations of time and reality found in racing photo-finish images. A photo finish depicts objects moving past a fixed vertical strip of space (the finish line) over time, using a moving strip of film which moves inside of a shutter-free camera at the estimated speed of the racing objects. In a photo finish, the usual expectation – that a photograph will reveal spatial relations at a given instant of time – is reversed. Instead, the resulting photograph depicts only moving objects as they pass a fixed location, in a vertical strip of recording field, over elapsed time. Photo finishes are regarded in a court of law as absolutely truthful to the order of race finishes. Yet the images are full of visual departures from known reality. This chapter will discuss the photo finish's complicated relationship to time, space, and "truthfulness." Acceptance of the photo finish's blending of realism with abstraction will be examined as a product of a mix of an enthusiasm for the rationalizing, positivist quest to explain the world, and an acknowledgement of the limits of its comprehension by humanity. Both of these urges will be discussed as necessary components of modernity's cycle of innovation, as knowledge of the parameters of our understanding of the universe is the very fuel that drives further technological advance.

In Chapter 5, an examination of the temporal complexities of the NASA's whole-Earth composite photographs shifts the conversation to include today's digital image-making practices. NASA's *Big Blue Marble Next Generation* images are made with the aid of software which blends photographs from many satellite images to produce a picture of what the Earth might look like if there were, for example, no clouds, and if the planet were bathed in full sunlight. Although these images are regarded as truthful scientific inscriptions and rarely are questioned, these whole-Earth photographs are the result of the seamless digital combination of separate photographs, all of which relate to *different* moments in time. What we see in them is a complex synthesis of fractured time, collapsed space, and simultaneity of vision. As Geoffrey Batchen has suggested, digital images are "*in* time but not *of* time."[3] These unique composite temporal abstractions are products of an anxious culture coming to grips with a quickening pace of life, a quickening facilitated by computer technology – the very tool used to compose these images. But they also share in photography's history as a medium for making data visualizations that test the expectations of documentary veracity often monolithically associated with photography.

This book self-consciously examines a selection of understudied, yet commonplace, photographic practices that appear in our homes, on the pages of our newspapers and magazines, at race tracks, the Tour de France, and at the Olympic Games, and in our textbooks. Though arguably unextraordinary, these images have a formative influence on our knowledge of the world and our sense of family identity. They are part of our human temporal habitat. By selecting these objects whose stories have not yet been told, this project contributes to the expanded discussion of the diverse field of the "photographic," while also undertaking the task of investigating a few of the medium's many and varied articulations of time.

The Multifariousness of Photographic Time

Despite the importance of the fluid expressions of time by the medium of photography, an expanded study of temporality, and its genesis in specific, distributed photographic micro-cultures, has not been undertaken in our existing scholarship. The bulk of the existing literature on photographic temporality and its cultural/intellectual genesis concentrates on the work of the pioneers of late 19th-century to early 20th-century motion studies. Notably, Marta Braun's scholarship on Étienne-Jules Marey offers valuable biographical and scientific contexts for these early motion-study projects.[4] Other scholars, including Philip Prodger, Rebecca Solnit, and François Dagognet, have offered similar biographical and contextual studies of Eadweard Muybridge.[5] A recent book by an historian of science, Jimena Canales, also takes a closer look at Enlightenment scientists' "tenth-of-a-second-minded" interests, and suggests that their work inspired Marey and Muybridge to capture the "microsecond."[6] While Canales's study is written from the perspective of an historian of science, it represents an interdisciplinary approach to the history of photography that accounts for the medium's connection to its own scientific and technological contexts. Even though the degree to which individual photographers such as Marey and Muybridge may or may not have been aware of such discourses is debatable, Canales's work to illustrate the merits of further interdisciplinary views of photographic practices have influenced the conception of this book. As these scholars have shown, studying the role of the discursive climate that gave form to photography's variegated expressions of facets of time, if localized to the image itself, holds much promise for illuminating and enriching our existing body of scholarly literature.

Given the wealth of scholarship on late 19th- to early 20th-century motion studies, photography is frequently discussed by scholars as offering a "stilled," fraction-of-a-second-exposure, "instantaneous-time" image that comprises a constituent building block for cinema.[7] As a result, many scholars have been content to articulate photography's temporal dimensions – to use Solnit's words – as the unique, "violent, abrupt, glorious" rupture offered by the instantaneous-time photograph.[8] As Solnit explains, photography's expression of a sliver of time is decidedly "modern," and expresses the "sudden shock" of living in "a transformed world" in which the absolutism of time and space had been "annihilated."[9] Mary Ann Doane has suggested that this "shock" was further accentuated by the monetization of time itself as a thing of value.[10] This shift was nothing short of revolutionary, and echoed then-contemporary, parallel scientific innovations in the fields of engineering and physics, as well in the new assembly lines suddenly appearing in full force in factories.

While acknowledging the important role of the late 19th-century photographic motion studies, this book complicates the assumed straightforward temporal expressions of "instantaneous" photography, and it offers a look at photographs from varying periods throughout the medium's history – including the digital age. It addresses more than only the "atomized" time encapsulated by the early motion studies of Marey and Muybridge. Rather, Chapter 2 takes a look at the temporal complexity of other overlooked photographic practices with their own rich stories, including the stroboscopic photograph (which achieves the "atomized" time sought by these pioneers), fin-de-siècle family portrait, racing photo finish, and NASA's digital-composite montages. As such, this book examines the temporal anxieties peculiar to the cultures of each of these modes of image-making, and it aspires to present historical narratives to supplement the literature concentrated on motion studies and their engagements of industrial and technological modernity. In particular, Chapter 2's case study of an apex of "atomized" time, Edgerton's stroboscopic images, extends the early motion-studies of Marey and Muybridge to the crowning achievement of the visual representation of the millionth-of-a-second-view of time. Yet it proves to be anything less than straightforward in its relation to both the "real," and to time itself – especially when viewed contextually.

This study also is premised on the established arguments that photography has more than one manner of expressing time, and it acknowledges the work of other scholars who have played an important role in these discourses. For example, Johanna Drucker has noted that "the myths of temporality specific to photography – the frozen moment, the window and mirror imagery, the language of the gaze, of the look, the glance – are all attached to the singularity of photographic time."[11] Each image, she suggests, materializes multiple "times": the "time of exposure, historical time, time of development, [and] cropping."[12] In addition to these innate material temporal investments present in a photograph, Drucker also notes the ramifications that the "time of reception and circulation" may have upon the image, although her study focuses on the video-portraiture work of artist Scott Kildall. Chapter 3 contributes to the exploration of a photographic time mentioned by Drucker, and offers one that exceeds that of solely the image's material production to illuminate the importance and essentially social functions of the often open-ended, "reception"-produced, narrative temporalities.

Jan Baetens and Hilde van de Gelder likewise have cautioned against considering photography's purview of time as a singular, monolithic entity.[13] They suggest that photographs instead be viewed as the subjects of "a complex social process" that uses them to "produce social meanings and interactions that evolve in time and through time."[14] This is to say, they encourage future scholars to view the meaning-production of photographs as a mobile, changeable entity, with respect to temporality, and innately bound to their respective cultures. This book explores that premise in detail – especially in Chapters 3 and 5.

Cinema and photography scholar Damian Sutton further poses the possibility that the photographic expressions of time are driven by human biology's clash with that of society's forced rhythms in an age of industrial modernity: "To photograph is to produce stress between societal clock time and biological time."[15] Therefore, the medium offers fundamental refuge, in Ulrich Baer's words, "to a time that is radically contingent," that is "not ours."[16] Photographic time, these scholars collectively suggest, is innately an "other" to our bodily experience of time itself, something Jacques Derrida has called "a heterogenous time" comprised of "differing/deferring and differentiated duration."[17] Derrida further notes that in a photograph, "artifices" occur as a result of this "differing/deferring" that prompt the image to "definitively break with presumed phenomenological naturalism."[18] In sum, these scholars turn to photography as a medium through which time's innate foreign-ness, "other-liness," and strangeness with respect to the experience of time with the unaided human senses may be explored. This book makes a contribution to that project.

Photography, as the product, embodiment and faithful servant of a complex and variegated modernity, is equally variously temporally stratified. Photography's story is inextricably bound to dialogues about its own technological advances, we well as those of the unique temporal cultures in which its various practices has been employed as an expressive medium. Even the most ordinary, commonplace sampling of photographs, this book suggests, possess the ability to represent many levels of the temporal experience. Photography's flexible expressions of time are both symptomatic of technological modernity and postmodernity as a general phenomenon, and is a manifestation of the photographic medium's conditional relationship to reality, a relationship that arguably has been complicated by shifts in photographic technology and media, including the medium's digitization. These trends are shaped by the medium's status as one among many technologies that redefined time and space, and – in turn – were shaped by the context that gave form to them. This is to say, the medium of photography is both a subject and catalyst of technological modernity and postmodernity.

Notes

1 Roland Barthes, *Camera Lucida* (New York: Hill and Wang, 1981).
2 Barthes, 115.
3 Geoffrey Batchen, *Each Wild Idea: Writing, Photography, History* (Cambridge, MA: MIT Press, 2001): 140.
4 Marta Braun, *Picturing Time: The Work of Étienne-Jules Marey (1830–1904)* (Chicago and London: University of Chicago Press, 1992). Braun studies additional photographers, including Eadweard Muybridge and Harold Edgerton, in these published essays: Marta Braun, "The Expanded Present: Photographing the Moment," in *Beauty of Another Order: Photography in Science*, Ann Thomas, ed. (New Haven: Yale University Press, 1997), 150–184; Marta Braun, "Time and Photography," in *Tempus Fugit: Time Flies*, Jan Schall, ed. (Kansas City, MO: Nelson-Atkins Museum of Art, 2000), 132–143. Martha Langford's study of family photography albums suggests that the performance of telling stories about those objects is an extension of oral tradition. She implies that the oral narration of a photographic album allows for departures from a linear narrative structure: Martha Langford, *Suspended Conversations: The Afterlife of Memory in Photographic Albums* (Montreal, London and Ithaca: McGill-Queen's University Press, 2001).
5 Phillip Prodger, *Time Stands Still: Muybridge and the Instantaneous Photography Movement* (Oxford: Oxford University Press, 2003); Rebecca Solnit, *River of Shadows: Eadweard Muybridge and the Technological Wild West* (Middlesex, England: Viking Penguin, 2003); François Dagognet, *Étienne-Jules Marey: A Passion for the Trace* (New York: Zone Books, 1992).
6 Jimena Canales, *A Tenth of a Second: A History* (Chicago: University of Chicago, 2009).
7 This conclusion underlies, for instance, these books by cinema scholars: Mary Ann Doane, *The Emergence of Cinematic Time: Modernity, Contingency, the Archive* (Cambridge, MA: Harvard University Press, 2002); Damian Sutton, *Photography, Cinema, Memory: The Crystal Image of Time* (Minneapolis: University of Minnesota Press, 2009).
8 Solnit, 81.
9 Solnit, 11, 81.
10 Mary Ann Doane, "Real Time: Instantaneity and the Photographic Imaginary," in *Stillness and Time: Photography and the Moving Image*, David Green and Joanna Lowry, eds. (Brighton: Photoworks/Photoforum, 2006), 23.
11 Johanna Drucker, "Temporal Photography," *Philosophy of Photography*, Vol. 1, No. 1 (2010): 23.
12 Drucker, 23.
13 Jan Baetens and Hilde van de Gelder, eds. *Photography and Time* (Leuven: Leuven University Press, 2010), viii.
14 Baetens and van de Gelder, viii.
15 Damian Sutton, "Delay: Notes on Photography as Non-Representational Thinking," in *Time Expanded* (Madrid: La Fábrica/PHotoEspaña, 2010), 149.
16 Ulrich Baer, "The Shutter's Suspension of Time: Toward a Democritean Theory of Photography," in *Time Expanded* (Madrid: La Fábrica/PHotoEspaña, 2010), 50–51.
17 Jacques Derrida, *Copy, Archive, Signature: A Conversation on Photography* (Stanford: Stanford University Press, 2010), 8.
18 Derrida, 8.

Bibliography

Baer, Ulrich. "The Shutter's Suspension of Time: Toward a Democritean Theory of Photography," in *Time Expanded*. Madrid: La Fábrica/PHotoEspaña, 2010. 43–51.

Baetens, Jan, and van de Gelder, Hilde, eds. *Photography and Time*. Leuven: Leuven University Press, 2010.

Barthes, Roland. *Camera Lucida*. New York: Hill and Wang, 1981.

Batchen, Geoffrey. *Each Wild Idea: Writing, Photography, History*. Cambridge, MA: MIT Press, 2001.

Belden-Adams, Kris. "Modern Time: Photography and Temporality," *International Journal of Technology, Knowledge, and Society*, Vol. 5, No. 3 (2009): 25–42. doi:10.18848/1832-3669/CGP/v05i03/55994.

Braun, Marta. *Picturing Time: The Work of Étienne-Jules Marey (1830–1904)*. Chicago and London: University of Chicago Press, 1992.

Braun, Marta. "Time and Photography," in *Tempus Fugit: Time Flies*, Jan Schall, ed. Kansas City, MO: Nelson-Atkins Museum of Art, 2000. 132–143.

Braun, Marta. "The Expanded Present: Photographing the Moment," in *Beauty of Another Order: Photography in Science*, Ann Thomas, ed. New Haven: Yale University Press, 1997. 150–184.

Canales, Jimena. *A Tenth of a Second: A History*. Chicago: University of Chicago, 2009.

Dagognet, François. *Étienne-Jules Marey: A Passion for the Trace*. New York: Zone Books, 1992.

Derrida, Jacques. *Copy, Archive, Signature: A Conversation on Photography*. Stanford: Stanford University Press, 2010.

Doane, Mary Ann. *The Emergence of Cinematic Time: Modernity, Contingency, the Archive*. Cambridge, MA: Harvard University Press, 2002.

Doane, Mary Ann. "Real Time: Instantaneity and the Photographic Imaginary," in *Stillness and Time: Photography and the Moving Image*, David Green and Joanna Lowry, eds. Brighton: Photoworks/Photoforum, 2006. 23–38.

Drucker, Johanna. "Temporal Photography," *Philosophy of Photography*, Vol. 1, No. 1 (2010): 22–28. doi:10.1386/pop.1.1.22/7.

Langford, Martha. *Suspended Conversations: The Afterlife of Memory in Photographic Albums*. Montreal, London and Ithaca: McGill-Queen's University Press, 2001.

Prodger, Phillip. *Time Stands Still: Muybridge and the Instantaneous Photography Movement*. Oxford: Oxford University Press, 2003.

Solnit, Rebecca. *River of Shadows: Eadweard Muybridge and the Technological Wild West*. Middlesex, England: Viking Penguin, 2003.

Sutton, Damian. *Photography, Cinema, Memory: The Crystal Image of Time*. Minneapolis: University of Minnesota Press, 2009.

Sutton, Damian. "Delay: Notes on Photography as Non-Representational Thinking," in *Time Expanded*. Madrid: La Fábrica/PHotoEspaña, 2010.

2 Photography, Instantaneity, and the "Frozen Moments"

After being lifted into the air by hurricane-force winds in an African desert, the protagonist of the French science-fiction novel *Giphantie: A View of What Has Passed, What Is Now Passing, and During the Present Century, What Will Pass in the World* is unconsciously whisked into a dark room on a lush, mystical island. With a governing spirit figure as his guide, the traveler is drawn to a window through which a raging storm appears. Surprised, because he thought the storm already had passed, the protagonist rushes to peek his head through the window to see if the tempest had resurged. The traveler is stunned as he hits his forehead on what is not a window, but a solid canvas panel bearing a stilled, highly detailed image of the storm. He staggers backward several steps, and his guide explains that the deceptively realistic, mirror-like image he sees frozen on that "window" panel covered in a mysterious, viscous material is the "elementary spirits'" successful attempt to "fix these transient images" "in the twinkle of an eye." The traveler stares with great awe at this visual manifestation of a frozen slice of time, halted and thereby "retained" for his later re-viewing.[1]

This novel, published originally in 1760 by Charles-François Tiphaigne de la Roche, offered what has been described by several historians and commentators as a prophetic description of the object that we have come to know as "a photograph": an image made by a gelatinous chemical concoction that creates a permanent, stilled, two-dimensional, visual likeness of a scene on a flat surface. The necessary scene-recording duration of the chemical goo's exposure, Tiphaigne de la Roche writes, is only a "twinkle of an eye," as is the strange view of time it offers. Nevertheless, one of the central qualities of this science-fictional proto-photograph is its unique capacity to express a frozen slice of estranged time that noticeably differs from the "transient" experience of "lived" time. It is familiar and an "other," simultaneously.

Giphantie's innately temporal definition of what we would come to know as "photography" propelled excerpts from this science-fictional book to the attention of members of photography organizations in England and the United States in the nineteenth century. It still is cited today in our histories as an illustration of a proto-photographic cultural fascination with the prospect of halting time and keeping a thin slice of it as a stilled, visual image. Decades after the announcement of the invention of photography, references to *Giphantie* appeared in historical narratives for the medium by daguerreotypists Ernest Mayer, Louis Frédéric Mayer, and Pierre Louis Pierson, and historians Josef Maria Eder, Georges Pontonniée, and Beaumont Newhall.[2] The novel has been cited and discussed in more recent texts by

Tanya Sheehan and Andres Zervigon, Geoffrey Batchen, Robert Hirsch, and Naomi Rosenblum.[3] *Giphantie*'s emphatic fascination with cheating the ongoing continuum of "lived" time by capturing a small atom of it as a stilled image was hardly unique to this prescient work of science-fictional proto-photography. It merely set the stage for a host of additional references to the medium's unique capacity to isolate a thin slice of time that appears liberally in the earliest accounts of its invention. *Giphantie* culturally contextualized a general fascination with temporality that was pervasive during the Romantic era during which the medium emerged.

The "Instant": A Romantic Era Visual-Culture Fascination

On the reverse of the oldest-surviving photograph, Joseph Nicéphore Niépce's *View From the Window at Le Gras* (c. 1826) (**Fig. 2.1**), Francis Bauer – who encouraged Niépce to write a letter staking claim to the invention to the Royal Society (of which he was a Fellow) – added an explanatory inscription in 1827. The "Heliography" process used by Niépce, Bauer testified, was one by which an image was acquired "spontaneously by the action of light."[4] Bauer's written remarks bear witness to Niépce's claim to have invented photography, while also noting the instantaneous speed with which Niépce's chemistry began translating light into an image. (Bauer was not present in person in Gras, France, to witness the rapidity of the photochemical reaction itself. He nonetheless imagines it, prompted by Niépce's description.)

The "spontaneity" to which Bauer testified differed significantly from the "instantaneous" exposures that predominate in today's practices of the medium,

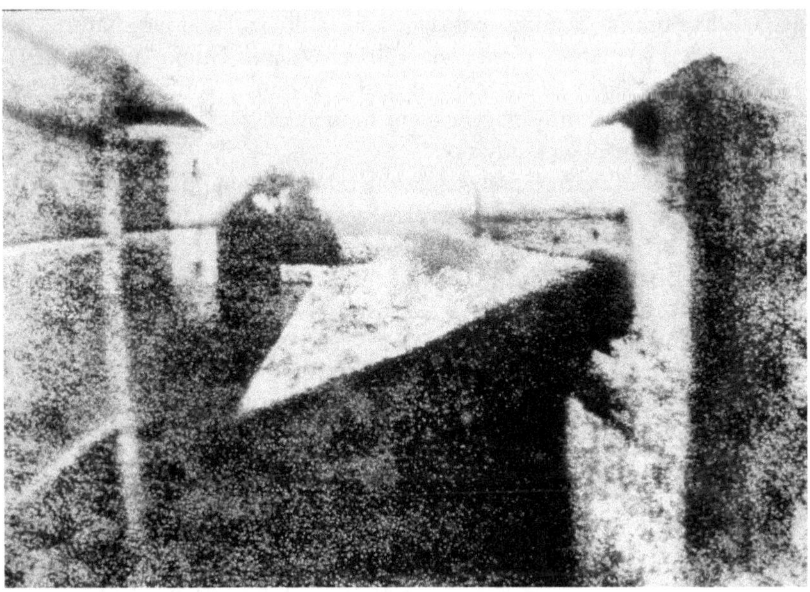

Figure 2.1 Joseph Nicéphore Niépce, *View From the Window at Le Gras*, c. 1826. Public Domain. (Original: 7.87 in. × 9.84 in./20 cm × 25 cm). Due to the eight-hour exposure, the buildings are illuminated by the sun from both the right and left.

which most commonly result from fraction-of-a-second slivers of time in a full field of space. Niépce's *View From the Window at Le Gras* resulted in about *eight hours* of exposure of the pewter plate to light. The sun even shifted to fully illuminate two opposite-direction-facing structures as the image was being made. Bauer's recognition of the photochemistry's "spontaneous" reactivity likely was both an expression of his awe for Niépce's achievement, and his communication of a different view of time, as it was typically invested in the creation of objects of visual culture. In 1826, naturalistic painting and printmaking were *the* measures of the necessary temporal investment (of at least *several weeks*) to create a realistic likeness of a scene, subject, or object with the standard degree of studied naturalism. Photography's promise of creating a lasting visual image in *only eight hours* was all the more radical, given these prevailing, painting-and-printmaking-based expectations of the temporal investment of visual-culture production in the early nineteenth century.

As Geoffrey Batchen and Jennifer Green-Lewis have suggested, a unique, keen awareness of time's passage emerged as a common subject of the poetry and painting of the Romantic period, which was a mainstream cultural fascination in the decades surrounding the invention of photography.[5] British landscape painter John Constable – working in England at the same time as pioneering photography inventor William Henry Fox Talbot – strove to arrest the transience of nature in his "six-footer" landscape paintings and prolific cloud studies.[6] Constable's incessant attention to capturing the fleeting qualities of meteorological phenomena reveals a unique Romantic compulsion for stilling the ephemeral (manifested in nature) as a lasting visual image. Batchen suggests that one of Constable's contemporaries, poet Samuel Taylor Coleridge, similarly sought to express the ephemeral in poetic form. Romanticism itself embodied a proto-photographic manifestation of the "desire to photograph" – a yearning to fix time's fleeting passage and still its fugitive nature.[7] Batchen also examines texts written by Coleridge, Talbot, William Gilpin, and Thomas Wedgwood, and compares them to the paintings of Constable, arguing that these are "as much a study of the temporal exigencies of human observation, and thus of human subjectivity, as a picture of a cloudy sky."[8]

While nature was a central preoccupation of the Romantic writers and painters such as Constable, many of the earliest pioneers of photography also expressed their unabashed wonderment at the sheer capacity for natural forces to enable such a method of image-making in the first place. Bauer's inscription for *View From the Window at Le Gras* further declares it "Monsieur Niépce's first successful experiment of fixing permanently the Image from Nature."[9] Niépce's Heliography thus captured an unchanging, two-dimensional, frozen manifestation of the dynamic phenomena of "Nature itself."

Daguerre, the Diorama, and Departures From "Lived-Time" and "Reality"

After failing to find royal patronage or interest in the scientific societies – including Bauer's – Niépce entered a partnership in 1829 with one of the few men who *was* interested in the innovation, renowned Diorama and theatrical-scene painter Louis Jacques Mandé Daguerre, who offered Niépce his "new combination of camera obscura, his talents, and his industry."[10] Even with Daguerre's connections and showmanship skills on his side, Niépce died in 1833, leaving his family relatively penniless, his photographic process imperfect, and all of his ideas about photochemistry in the hands of a man who made measurably fewer contributions to their partnership than he did.

As heir to Niépce's experiments, Daguerre knew how to capitalize on the public's fascination with all things temporal. Along with Charles-Marie Bouton (a student of French Neoclassical painter Jacques-Louis David), Daguerre invented the Diorama on July 11, 1822, in Paris.[11] Dioramas were popular theatrical exhibitions of two double-sided, transparent paintings – each 72 feet wide and 45 feet high – that were viewed by an audience of about 350 people who sat on moveable seats and viewed the paintings under color-changing spotlight effects planned by Daguerre. Diorama shows featured imagery depicting typical Romantic themes: ruined cathedrals, idyllic Italian landscapes, and quaint mountain hamlets. Daguerre was particularly well-known for his Diorama-scene-painting prowess. These shows conjured the illusion of time's passing by changing the color and intensity of lighting in front and behind the painting, and by adding and adjusting various shades and colored filters attached to the spotlights. Daguerre often evoked the temporal shift between night and day by using a process he called "decomposition of form," which allowed him to make some objects appear, while others disappeared. As he explained, "if a green and a red part of the painting are illuminated by red light, the red object will vanish while the green one will appear black, and *vice versa*."[12] Daguerre's strategic combination of opaque and translucent paint, transmitted and reflected light, and his use of colored screens created a great range of scenic effects that dazzled audiences by allowing them to see a view of a compressed passage of time that was otherwise too long in duration to be dynamically accessible to the human eye.

For example, Daguerre's *Night Scene Diorama Painting of Alpine Scene With Chalet* (1836) (**Fig. 2.2**) reveals a quiet, idyllic, Swiss Alpine village slowly transitioning to slumber. Within the span of fifteen minutes, Daguerre shifted lighting effects to enhance certain parts of the painting at the expense of the visibility of others. This created the illusion of an entire night's passage, compressed into a quarter of an hour. Viewers then beheld the results of the avalanche in the year 1820 that buried a cluster of houses in the background, as the inhabitants of Alagna were said to have been "awakened to sleep – forever!" and Daguerre's *Morning Scene Diorama Painting of Alpine Scene With Chalet* (1836) (**Fig. 2.3**) appeared.[13] Time's passage was compressed and controlled, as a twelve-hour span from dawn to dusk transpired before the audience in just *fifteen tragic minutes*. Daguerre's naturalistic paintings entranced audiences with uniquely elastic, compressed views of time, while also testifying to a Romanticist theme of nature's vast, ultimate power over the plight of humankind. The careful orchestration of varied light effects upon the painted image, however, can be seen as an expression of humanity's domain over nature, much like the earliest successful photographs worked to manage "nature's" agent of light.

Although Daguerre had not succeeded in independently producing a single fixed, or even unfixed, photograph by the time Niépce died, Daguerre turned his attention to photography after declaring bankruptcy (with a debt of 60,000 francs – or $298,971.53 in United States dollars in 2017[14]) after the Diorama struggled to make ends meet through the 1832 cholera epidemic. Daguerre's first successful attempts to permanently "fix" pictures date to 1837. To secure funding for his

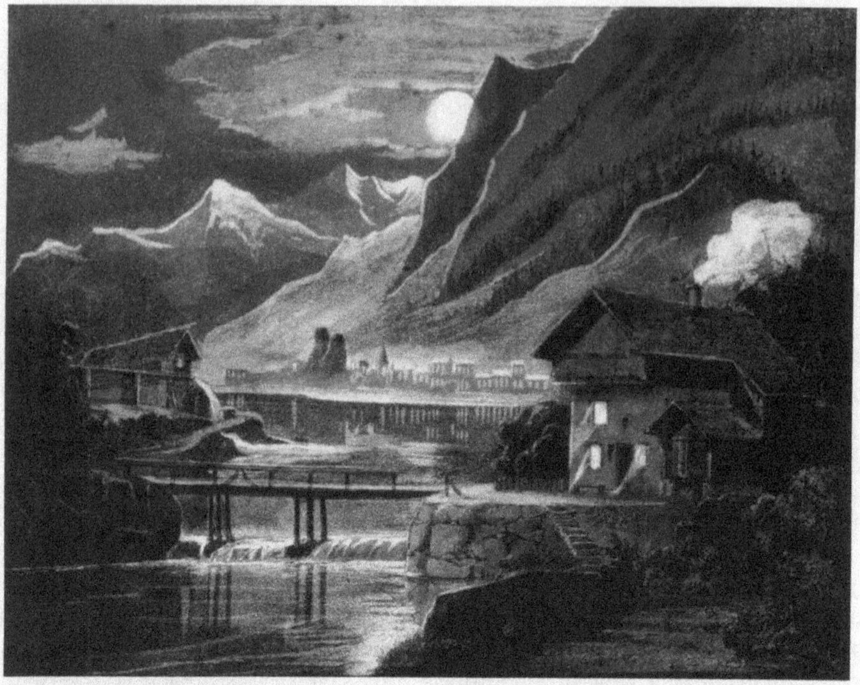

Figure 2.2 Louis Jacques Mandé Daguerre, *Night Scene, Diorama Painting of Alpine Scene With Chalet*, 1836. Copy print. Helmut and Alison Gernsheim Papers/Harry Ransom Center – The University of Texas at Austin.

photographic process, Daguerre stirred up great publicity by touring Paris with a camera on wagon, making photographs of monuments and public buildings along the way. Although he failed to find enough buyers of shares in the prospect of his invention to edge out Niépce's relatives from having cause to make claim to his work, he approached François Dominique Arago, an influential ally with connections to the French government and the Académie des Sciences. The distinguished physicist Arago agreed to test and present Daguerre's invention to the Académie at its meeting on January 7, 1839. This meeting featured a description of Daguerre's photochemistry, and Arago's introduction of the daguerreotype impressed and dumbfounded many audience members. Many rushed out and bought materials to emulate the process themselves. The French government purchased the daguerreotype process and the intellectual property rights to the Diorama. The "daguerreotype" was announced to the world and attributed in namesake to only Daguerre – although Niépce's son and widow received a small portion of the proceeds. The French expected nothing in exchange for sharing the process with the public but admiration, and the claim that photography was a *French* invention. Daguerre instantly became a celebrity and toured Paris to give public demonstrations of his process.

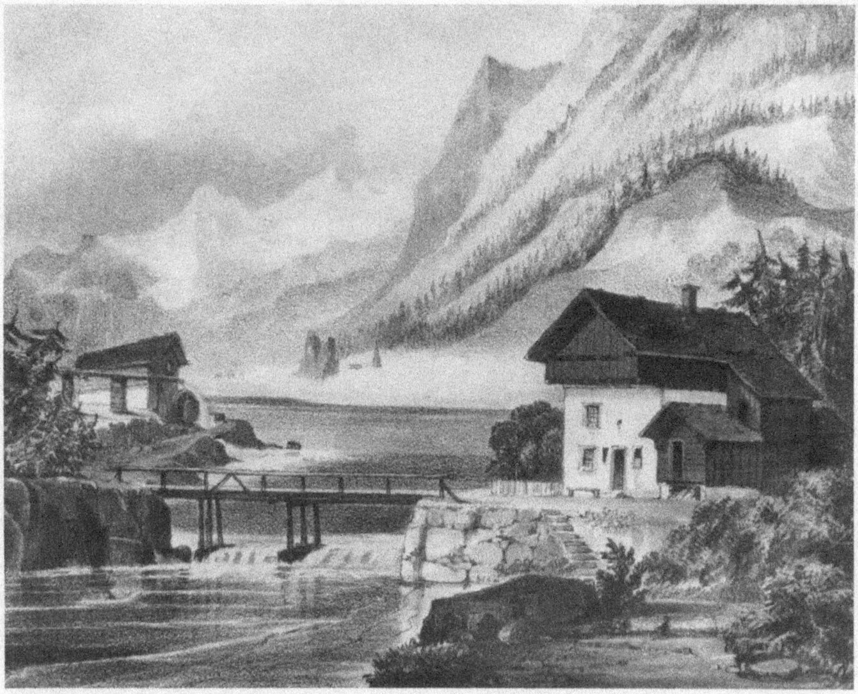

Figure 2.3 Louis Jacques Mandé Daguerre, *Morning Scene, Diorama Painting of Alpine Scene With Chalet*, 1836. Copy print. Helmut and Alison Gernsheim Papers/Harry Ransom Center – The University of Texas at Austin.

Arago's presentation and Daguerre's exhibitions kindled conversations about the peculiar views of time made possible by photography. On the day before Daguerre's invention of a method to chemically fix and still the projection of light through a camera obscura was presented by Arago to the Académie, journalist Hippolyte Gaucheraud "scooped" the announcement with an article that appeared in the January 6, 1839, issue *of La Gazette de France* (Paris).[15] In the article, Gaucheraud identified photography's unique and revolutionary relationship to time and temporal portability as one that "upsets all scientific theories of light and optics" by offering images that "are no longer transient reflections of objects, but their fixed and everlasting impress which, like a painting or engraving, can be taken away from the presence of the objects."[16] Here, Gaucheraud notes three peculiar dimensions of photographic time: the medium's revolutionary capacity to still an ephemeral likeness of a subject, to lend that image the possibility of eternal existence, and to be transportable for viewing and re-viewing by viewers across various spaces and times.

A daguerreotype's representation of time is yet more complex than that, the American inventor Samuel B. Morse suggested, in his written commentary after viewing Daguerre's *View of the Boulevard du Temple, Paris* (1839) (**Fig. 2.4**):[17]

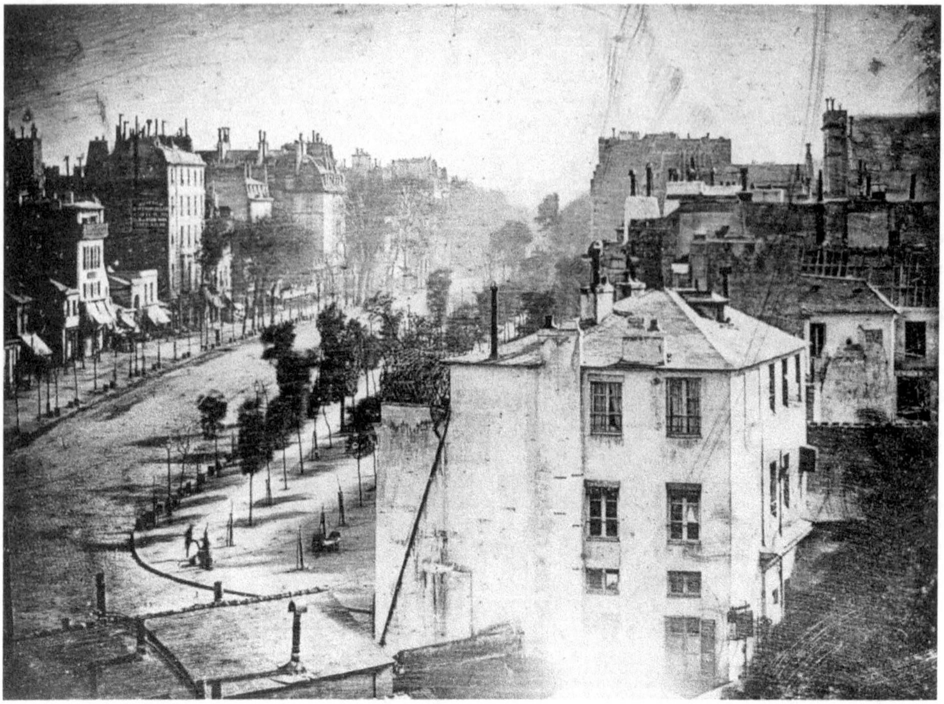

Figure 2.4 Louis Jacques Mandé Daguerre, *View of the Boulevard du Temple, Paris*, 1839. (Public Domain)

> Objects moving are not impressed. The Boulevard, so constantly filled with a moving throng of pedestrians and carriages, was perfectly solitary, except an individual who was having his boots brushed. His feet were compelled, of course, to be stationary for some time, one being on the box of the boot-black, and the other on the ground. Consequently, his boots and legs are well defined, but he is without body or head, because these were in motion.[18]

Other viewers, including Gaucheraud, were quick to point out that the unreal realism of Daguerre's street scene was due to the camera's extended recording of time's passage. Gaucheraud notes that

> all that was walking or moving does not appear in the picture; of the two horses of a hackney coach on the stand, one unfortunately moved its head during the exposure so the animal appears without a head in the picture.[19]

By explaining photography's peculiar departures from the appearance of "lived time" and the ongoing motion and temporal duration of an animate world, both Morse and Gaucheraud point out the strangeness of photography's representations of stilled time. They also offer a succinct expression of expectations of photography's relationship to time while also issuing it an implicit challenge to better capture motion and realism by reducing exposure durations.

Talbot's "Fairy Pictures": "Fixing Shadows," the "Momentary"

After Daguerre's innovation was introduced by Gaucheraud's article and Arago's presentation to the Académie, British newspapers spread news of Daguerre's groundbreaking invention. The first article to mention his work hailed him as a "prodigy," much to the certain chagrin of English scientist and linguist William Henry Fox Talbot, who had been working simultaneously and independently, without formal acknowledgement, on his own photographic process since, he claimed, 1834.[20] In *The Pencil of Nature*, Talbot describes the genesis of his initial curiosity with the prospect of photography as wholly temporal. After several failed attempts to sketch Lake Como in 1833, Talbot writes that this led me to reflect on the inimitable beauty of Nature's painting . . . fairy pictures, creations of a moment, and destined as rapidly to fade away."[21] He muses that "the idea occurred to me – how charming it would be if it were possible to cause these natural images to imprint themselves durably and remain fixed upon paper![22]

In the treatise "Some Account of the Art of Photographic Drawing, Or, the Process by Which Natural Objects May Be Made to Delineate Themselves Without the Aid of the Artist's Pencil" (1839), Talbot addressed not only the evolution of his photochemical experiments (which were inspired by the ideas and experiments of fellow Englishmen Thomas Wedgwood and Sir Humphry Davy), but he provided a philosophical view of the medium's relationship to time, too. Talbot referred to his early photographic experiments as "the art of fixing a shadow." Shadows, Talbot wrote, were "the most transitory of things . . . the proverbial emblem of all that is fleeting and momentary." He marveled at photography's ability to fix a shadow "for ever [sic] in the position which it seemed only destined for a single instant to occupy."[23]

While Talbot was not the first or the only to comment about photography's peculiar ability to render fleeting subjects eternal, his introduction of the photographic negative enabled multiple reproductions to be made from a single source image, further enabling the phenomenon of transportable, re-viewable, reinterpretive time that also was articulated with wonder in Gaucheraud's commentary on Daguerre's *View of the Boulevard du Temple*. Talbot thus would free the photograph from enjoying only a one-of-a-kind presence, and instead endowed it with the capacity for presence across multiple space-time nexuses.

Michael Sand, who has written on Talbot's photographic work, has postulated that the medium's unique ability to traverse, materially, into different points in the future may be explained, in part, by its context and status as a product of the late Industrial Revolution. The medium of photography was one of many time-reorienting new processes and inventions that sprang into being during this period of profuse innovation. In addition to its promise to "preserve the past and participate in the industrial present," as Sand suggested, photography also enabled the present to potentially communicate itself to the future.[24] This activates for photography the possibilities of unstable time relationships that lie outside of the material conditions of producing the image itself, in which the image is subjected to viewer-determination and reinterpretation. Chapter 3 explores photography's multifarious temporal condition, as viewer-determined "narrative times" that enable images to enjoy "time travel," and a lifespan that potentially exceeds that of its subjects.

As several of the earliest commentators on the medium have noted, photography emerged as a revolutionary medium of the Industrial Revolution that called for a re-examination of scientific theories of light and optics.[25] It arose in visual cultures in which naturalistic views of subjects and objects were only possible through the relatively prolonged, temporal investments of creating a naturalistic painting or preparing a plate for printing. Photography appeared before a public that was fascinated with the distilled, dramatic passage of time that lapsed quickly in Daguerre's dioramas, and that was preoccupied with the urgency of capturing fleeting time in the poetry of Coleridge, Percy Bysshe Shelley, and William Wordsworth. This same public also was consumed with then-popular trends in painting, including the sentimentally frozen, fugitive time that was poised on the precipice of industrialization and celebrated in Constable's Romantic landscape paintings (not to mention in the frozen dramatic moments of Joseph Mallord William Turner, Thomas Cole, and other then-contemporary Romantic painters). Unsurprisingly, the early commentators on the medium noted the new medium's peculiar manner of capturing a small fugitive slice of time, relatively "spontaneously," and fixing this strange image "for ever [sic]," in fitting with the temporal culture of the first few decades of the nineteenth century.

The "Instant," Canonized: Photography's Technological Evolution

Before long, a narrative thread based on these early founders' accounts of their experiments with the medium began to emerge. Temporality was an essential component of this history. Beginning with the announcement of the invention of photography, the medium's earliest practitioners almost immediately began a quest to improve upon their processes and reduce exposure times through improving the sensitivity of photochemistry and lenses and, eventually, rapid-moving shutters and spark/strobe illumination. In their comments about Daguerre's *View of the Boulevard du Temple, Paris*, Morse and Gaucheraud perhaps unintentionally lay out a technological challenge for the medium that becomes the narrative of many histories of photography: achieving an increased fidelity to the real by reducing exposure-time durations. Daguerre's fifteen- to thirty-minute exposures in 1839 were improved upon by the twenty- to ninety-second exposures enabled by "*quick-stuff*" (1841). Improved lens designs further decreased exposures to – depending on light quality – anywhere from ten to sixty seconds in duration.[26] Talbot's calotype process enabled photographers to make portraits with exposure times as brief as one or two minutes (1841).

Later processes continued this trajectory and enabled even faster exposures. Wet-collodion photographs taken outdoors reduced exposures to between one-half and five seconds (1851). Ambrotypes featured an exposure of two to twenty seconds (1851). But the introduction of gelatin emulsion in 1878 allowed exposures as brief as 1/25th of a second, which further was decreased to 1/250th of a second by 1900. Photography histories laud the photochemical sensitivity of gelatin emulsion, which also was celebrated in the literature of photography journals and camera clubs.[27] To capture that enthusiasm, Beaumont Newhall – photography historian/curator, and author of the influential *The History of Photography, From 1839 to the*

Present – quotes a commonly recited popular poem from 1881 by Marc Oute, which rejoiced in the light-recording speed of gelatin emulsion: "Collodion – slow old fogey! – your palmy days have been, You must give place in the future to the plates of Gelatine."[28] As Newhall also notes, "[t]he perfection of gelatin emulsion not only led to the conquest, analysis, and synthesis of action." The standardization of recommended exposure- and processing-time durations gave photographers the feeling of confident technical mastery over the production and reproduction of their visual records of time's fleeting passage by the late nineteenth century.

Light-sensitive chemistry and improved lenses quickly and aptly rose to the challenge of capturing Talbot's "creations of the moment."[29] In the eighth chapter of the book, titled "The Conquest of Action," Newhall introduces another domain of the medium which awaited technical mastery, and found it by the close of the nineteenth century: camera shutter speeds. He presents a story of the development of the fraction-of-a-second shutter speed, a narrative that takes readers on an eight-page overview of the experiments of Eadweard Muybridge, Étienne-Jules Marey, Thomas Eakins, and Ottomar Anschutz.[30] Muybridge, Newhall notes, invented one of the first camera shutters in 1869. Prior to that, lens covers were removed by hand, in a less-than-instantaneous fashion.[31] Newhall makes reference to the caption of one of Muybridge's early Palo Alto motion studies, *The Horse in Motion* (1878), and marvels at his achievement of shutter speeds of less than 1/2000th of a second.[32] With this development, as the title of his chapter implies, the technical challenge of mastering the representation of motion was "conquered," and so was the task of representing an infinitesimal slice of time:

> Now the increased sensitivity of negative material made it necessary to slice time into precise fractions of a second. Shutters of the most ingenious kind were designed in great variety . . . By the end of the century accurate exposures of 1/5000th of a second could be made.[33]

Once "conquered," the representation of a brief instant, or "atomized," view of time, in a full field, was institutionalized around the fin de siècle as "*the*" way a photograph relates to time. Eastman Kodak's box cameras, introduced in 1888 and marketed to the masses, featured exposure times of 1/20th of a second.[34] Box and Brownie cameras established for a mass-culture audience of picture-takers and viewers the normative expectation that vernacular "snapshot" photography – accessible to a broad public – operates by the logic of the temporal fragment. In the twentieth century, further photochemical and technical developments enabled for the conquest of even the picosecond (one-trillionth of a second) in the hands of Harold Edgerton. Also, as Chapters 5 and 6 will discuss in far more detail, digital technology introduced tools that also dramatically have shaped the temporal considerations inherent to the taking, creation, editing, and dissemination of images. As such, technology continues to have a significant influence over late twentieth- and early twenty-first-century photographic practices.

Photography as Catalyst and Symptom of Modernity

The first seventy years of the medium's history coalesce to create a narrative of improvements in photochemistry, camera shutters, and lens technology. These changes enabled the medium to eventually rise to a destiny of interrupting "the stream of time" so that "motion could be broken down into its constituent parts," as historian Miles Orvell has suggested.[35] By virtue of this, photography in effect engaged a challenge that had been defined by commentator Charles Baudelaire, who asserted that representing the contingency of everyday life in eternal form was a task that defined the very age of "modernity":

> The aim for him [the artist] is to extract from fashion the poetry that resides in its historical envelope, to distil the eternal from the transitory ... By "modernity" I mean the transitory, the fleeting, the contingent, the half of art whose other half is the eternal and the immutable.[36]

Our existing literature on photographic time typically focuses on the apexes of this journey toward achieving exposures of "thin slices of time," associating this with a uniquely "modern" urge to still a vital, atomized bit of time and render its representation eternal. One of those peaks occurred in the late nineteenth century with the motion studies of Muybridge (1870s and 1880s), and Marey (1880s–1904). These photographers' contributions to the task of revealing the appearance of objects during "thin slices of time" – appearances that were previously inaccessible to the unaided human eye – have been much celebrated in photographic scholarship.[37] Advances in photosensitive chemistry, combined with shutter-mechanism innovations, made exposure times of 1/1,000th of a second possible for the first time, at least under ideal lighting conditions.[38] The camera's sensitivity finally surpassed that of the human eye-to-brain relay response of 1/10th of a second by about one-hundredfold, offering viewers a radical visual abstraction from their biological experience of time.[39] Instead, observers were introduced to a unique manner of seeing the world "photographically."

The medium's revolutionary new temporal views of the world were not always embraced amenably by its audiences. Rebecca Solnit describes the unique, "violent, abrupt, glorious" rupture offered to late nineteenth-century viewers by the instantaneous-time photograph.[40] As Solnit explains, photography's expression of a sliver of time is decidedly "modern," and expresses the "sudden shock" of living in "a transformed world" in which the absolutism of time and space had been "annihilated."[41] Considering the centrality of the fraction-of-a-second-exposure photograph to our conceptions of the medium today, it is remarkably easy to forget how shocking photographic views of time were when they first were received by an audience.

Alternatively, while the "shock" of photograph's estranged views of time may be one aspect of its "modernity," it also marked a measured degree of human conquest over time itself. This is to say, time had been brought under the purview of human expression, and rendered fluid, expressive, and even sculptural as both subject matter and means of production. With the emergence of photography, the "humanization of time," as Guy Debord mused, might not only remain the domain of the mind, but a material object that occupied the external world.[42]

Spark and Strobe Illumination: Pushing the Boundaries of the Temporal Fragment

Viewers of *View of the Boulevard du Temple, Paris*, including Morse and Gaucheraud, noted with wonder the ways in which Daguerre's street views differed from the actual view of that street. But as exposure durations grew even more brief than the human eye-to-brain relay response of 1/10th of a second, the views captured by the photograph became unverifiable by the unaided human eye, and departed, again, from the "lived time" experience. This particularly was apparent in spark- and strobe-illuminated photographic practices, which emerged almost as early as the medium itself. While working alone in Michael Faraday's basement laboratory at London's Royal Institution in 1851, Talbot attached a page of the *London Times* newspaper to a wheel that was rotated in front of a camera containing one of his new and highly sensitive Amphitype plates in a darkened room.[43] As the wheel turned, Talbot exposed a few square inches of the newspaper page for about 1/2,000th of a second, using spark illumination from Leyden jars and a mirror.[44] The printed letters in the photograph, Talbot reported in a note to Faraday, were "just as sharp as if the disk had been motionless."[45] The image has not survived, and as Talbot scholar Larry Schaaf suggests, some technical details about Talbot's experiment were unclear.[46] Accounts of Talbot's experimentation with spark-illuminated photography appear in his own journals, and are mentioned in various mid-nineteenth-century scientific and photographic journals.[47]

About four decades later, two British physicists experimented with similar techniques to Talbot's. Charles Vernon Boys photographed bullets in mid-flight at 1,400 miles per hour in 1893.[48] And in 1894, Arthur Mason Worthington captured a crisp image of a splash of a drop of milk by discharging spark illumination from a bank of Leyden jars to achieve a ten-millionths-of-a-second exposure (**Fig. 2.5**).[49] But it was not until Massachusetts Institute of Technology electrical engineering graduate student Harold Edgerton started becoming interested in these earlier techniques that these spark-illumination experiments using even-more-rapid strobe lighting received sustained attention. In 1931, Edgerton invented a stroboscopic flash device that emitted instantaneous bursts of light. He created photographs in a pitch-black room, using a camera with an open shutter and flashes of light that were one microsecond – or 1/1,000,000th of a second – long. Edgerton's work affirmed a lesson Talbot learned in his Leyden jar experiments: that the speed of photography was no longer dictated by the speed of shutters (for they can be done away with), but rather by the much more rapid speed of light.

Edgerton's stroboscopy has been lauded as a technological milestone in the medium's mastery of the temporal fragment. James R. Killian Jr. explains this photographic practice's contribution in a passage from a book about Edgerton's work:

> Beyond the time threshold of human vision lies a world of rapid motion which can be seen only with the aid of accessories that manipulate time as the microscope or telescope manipulates space. The unaided eye cannot see the beautiful coronet-like splashes of milk drops falling in a pool, the wings of a hovering hummingbird, or the effects of a bullet in flight. Not until the development of high-speed or single-flash photography (except by the use of sparks) did it become possible to see these rapid phenomena or visually to harness time to space.[50]

As Killian notes, Edgerton's stroboscopic images of accessible, familiar objects – including milk drops, hummingbirds in flight, golf clubs and tennis rackets making contact with balls, water drops and streams, shock waves, bullets in

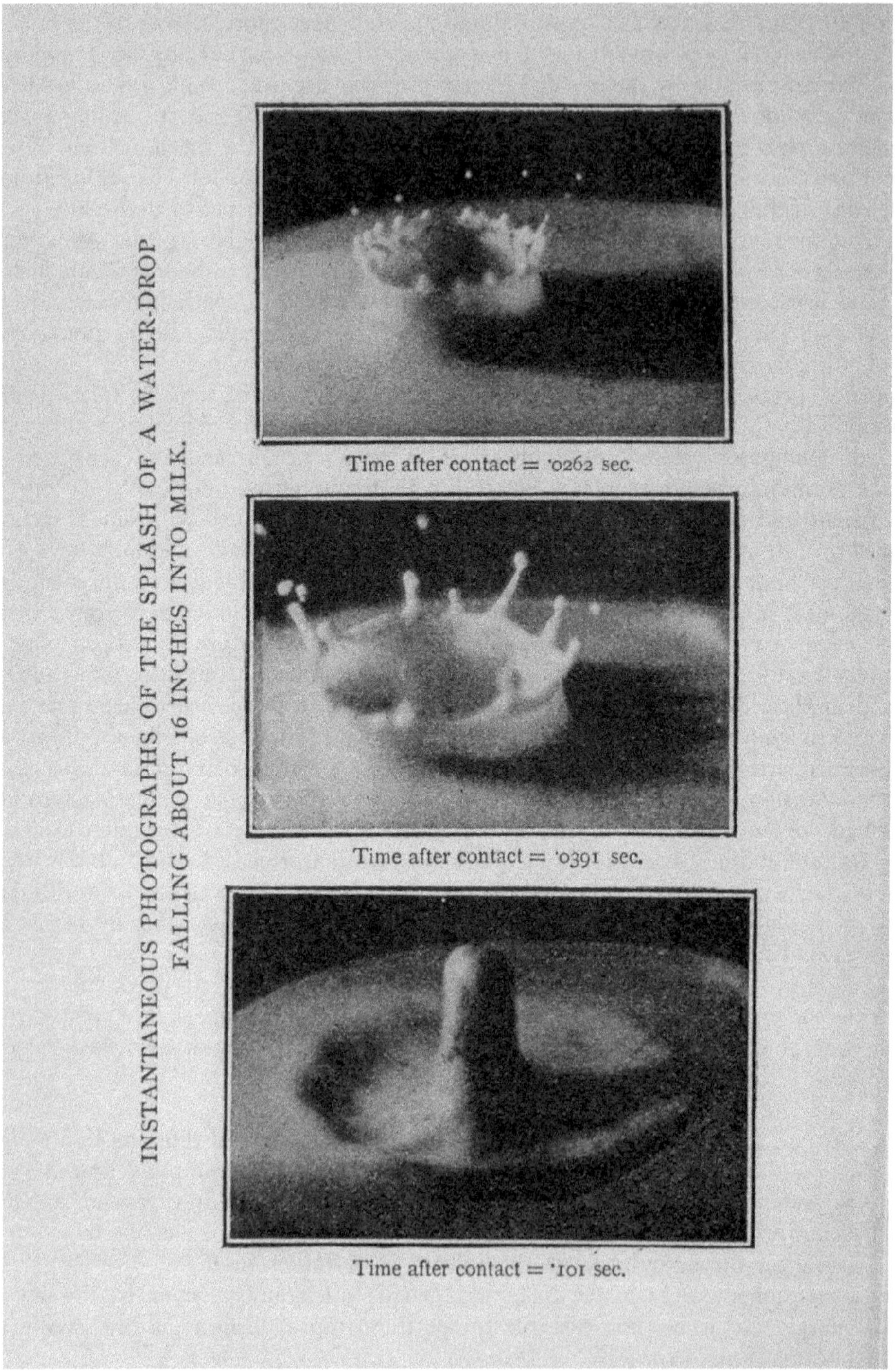

Figure 2.5 Arthur Mason Worthington, *Instantaneous Photographs of the Splash of a Water Drop Falling About 16 Inches Into Milk*, 1894. Reproduced in: A. M. Worthington, *The Romance of Science: The Splash of a Drop*, London: Society for Promoting Christian Knowledge, 1895. Reprint of the report delivered to the Royal Institution of Great Britain on May 18, 1894. (Public Domain)

motion, and dancers – are generally regarded as truthful to the appearance of objects in a full field of space during a millionth-of-a-second time exposure, even if these images offer blatant departures from the known appearance of the same objects in "lived" time.

Edgerton's representation of everyday subject matter – rather than of objects that might have appeal within the scholarly field of engineering – made his work accessible to a non-specialist, mass audience. In 1932, Edgerton's photographs were reproduced in trade publications such as *Electronics* and *Review of Scientific Instruments*, and a representative collection of his images was printed in successive months in the mainstream magazine *Technology Review*.[51] News of Edgerton's work quickly spread beyond the scientific community within just a couple years of his successful early experiments. In 1933, three of his photographs were exhibited for the first time in the Royal Photographic Society's annual exhibition in London, and his work was published in the popular periodical *American Golfer*.[52] In 1934, his life and work were presented in *American Magazine*.[53] Two years later, Edgerton was profiled in *Fortune* magazine.[54] But Edgerton's stroboscopic photographs became most well-known as part of a recurrent feature, "Speaking of Pictures," in *Life* magazine from 1936 to 1941.[55] This feature offered viewers a collection of Edgerton's stroboscopic photographs, explained by some brief introductory text and captions. "Speaking of Pictures" appeared near the front of each issue, where reader attention was the highest. In 1937, one of Edgerton's milk-drop photographs, *Milk Drop, Coronet*, was included in the Museum of Modern Art's seminal "Photography 1839–1937" exhibition, organized by Newhall.

Edgerton's peer and close friend, the photographer Gjon Mili, encouraged him to share his stroboscopic photographs with editors at *Life* magazine, which then semi-regularly featured a spread of his stroboscopic photographs. Particularly well-received were the reproductions of Edgerton's photographs of the breaking of light bulbs and of a bat and ball at impact that appeared in the November 20, 1939, issue of *Life*.[56] In the same year, Edgerton released a book of his stroboscopic images, *Flash! Seeing the Unseen by Ultra High Speed Photography*, which received critical acclaim.[57] At the request of the Kodak Corporation, Edgerton set up a booth at the World's Fair in New York City that featured a baseball-shooting cannon and allowed visitors to take their own stroboscopic pictures with it.[58] In just eight years after he introduced the stroboscope (1931), Edgerton's photographs not only became the subject of mass-cultural fascination, but he quickly became a widely respected scientist, an exhibiting photographer, and an author.[59]

The mass publication of Edgerton's work was not without precedent. In the mid-nineteenth century, contact prints of botanical specimens were translated into woodcut sketches and reproduced in magazines. Muybridge's animal locomotion studies also were disseminated in mainstream magazines and horse-racing periodicals as well as science journals. His photographs particularly kindled discussions among the general public about the visual departures photography offered from the experience of lived time. But the enthusiastic reception of Muybridge's images suggests mass culture's predominantly warm embrace of a "photographic" manner of seeing. Specifically, the representation of spatial relationships within a fraction of a moment of time became *the* manner by which the medium was thought to convey time. The invention of X-ray images was similarly announced immediately in mainstream magazines and newspapers. Readers were fascinated by the publication of early images of the "Röntgen Rays," and created poems, cartoons, and songs in tribute to the X-ray.[60] The X-ray, like Edgerton's stroboscopic images, offered a photographic glimpse of a reality inaccessible to the unaided human eye. Edgerton's photographs likewise were received with great enthusiasm and wonder. The images, Killian explains, particularly appealed to the public's thirst for seeing everyday objects engaged in "phenomena never before visible."[61]

Photography historian Marta Braun also notes that Edgerton's views of familiar objects frozen at 1,000,000th of a second collapsed "the intellectual boundary between entertainment and twentieth-century science," making science accessible to the masses.[62] The eager reception of images that complicated the public's conceptions of both truth and temporality in the late nineteenth and early twentieth centuries, cinema scholar Mary Ann Doane has suggested, is an indication of a shift in consciousness, one "specific to modernity and intimately allied with its new technologies of representation (photography, film, phonography). Time was indeed felt – as a weight, as a source of anxiety, and as an acutely pressing problem of representation."[63] Specifically, the workplace's new practices of "clocking in," "scientific management" trends including Frederick W. Taylor's "efficiency studies" (a practice in which photography was engaged as a data-collection method – especially at the hands of Frank and Lillian Gilbreth), and the development of the assembly line all strove to eliminate all unproductive time from the workplace. Taylor outlined "scientific management" as follows:

> The managers assume, for instance, the burden of gathering together all of the traditional knowledge which in the past has been possessed by the workmen and then of classifying, tabulating, and reducing this knowledge to rules, laws, and formulæ which are immensely helpful to the workmen in doing their daily work. In addition to developing a *science* in this way, the management take on three other types of duties which involve new and heavy burdens for themselves. These new duties are grouped under four heads: *First.* They develop a science for each element of a man's work, which replaces the old rule-of-thumb method. *Second.* They scientifically select and then train, teach, and develop the workman, whereas in the past he chose his own work and trained himself as best he could. *Third.* They heartily cooperate with the men so as to insure all of the work being done in accordance with the principles of the science which has been developed. *Fourth.* There is an almost equal division of the work and the responsibility between the management and the workmen. The management take over all work for which they are better fitted than the workmen, while in the past almost all of the work and the greater part of the responsibility were thrown upon the men. It is this combination of the initiative of the workmen, coupled with the new types of work done by the management, that makes scientific management so much more efficient than the old plan.[64]

Human beings were encouraged to operate as efficiently as machines, in accord with the commonly held belief across multiple industries that "time is money." Doane notes that "the desire to analyze and to rationalize time was frequently embodied as a desire to make time visualizable" and tangible, and to try to repossess some of the autonomy the worker had lost.[65] After all, keeping up with this new, precious, productive concept of capitalist time placed considerable stress on the human body, but was otherwise difficult to express visually – and was even more difficult to express phenomenologically. Workers came to notice that those who owned time's rigorous visualizations (managers and business owners) possessed power and privilege in capitalist industrial societies. They controlled the gaze, while the working caste became its surveilled subject. Several scholars have argued that "repossessing" time is a central struggle of the modern working class, especially in the first half of the twentieth century. This struggle was expressed in visual culture, as Charlie Chaplin struggles against the unrelenting cadence of the assembly line in the film *Modern Times* (1936), and in the "Job Switch" episode from 1952 of *I Love Lucy*, in which characters Lucy and Ethel go to work in a candy factory while their spouses stay home and do housework. To keep pace with production, the women begin eating candies and hiding them in their pockets and hats, prompting an impressed assembly-line manager to speed up the conveyor belt further. These expressions voiced mass-cultural pushback in response to what some saw as the overly rationalized, workforce-punitive, overly quantified thrust of scientific workplace studies and assembly lines. As the main visual medium for collecting data about inefficient and efficient production methods for workplace-efficiency studies, photography was a complicit partner in this enterprise of commodifying time.

Death of a (Many) *Light Bulb*(s): Examining Edgerton's Altered Series

But even within the discourse of science, less rigor existed than one might anticipate. Edgerton's multiple-stroboscopic-image series also offer another, more complex set of possibilities for exploring the temporal dimensions of stroboscopic photographs. *Death of a Light Bulb/.30 Caliber Bullet* (1936) (**Fig. 2.6**) consists of four photographs, arranged in two rows of two images each. The title of the photographic series, *Death of a Light Bulb/.30 Caliber Bullet*, provides the framework within which the meaning of these images unfolds. When "read" in sequential order (from left to right, top to bottom), the four photographs represent the sequential passage of a bullet through a light bulb, broken into presumably equal intervals of time. In other words, Edgerton leads his viewers to follow a "plot" of "death" from its inevitability (implied by an approaching bullet) in the first frame to the slow entry and internalization of the bullet by the bulb in the next two frames. In the final photograph, the bullet exits its victim, leaving a spattering of dust and a cracked, soon-to-collapse bulb in its wake.

A narrative "reading" of *Death of a Light Bulb* depends upon viewers to associate the images with each other, in linear order, and to imagine the unseen events that transpired within the durations left unseen, that transpired between these images. Images such as *Death of a Light Bulb* – which unfold sequentially under a title that imposes a "plot," like comic strips – create meaning by what Roland Barthes suggests is a system of "relay." He explains:

> Here (in cartoons and comic strips) text (a snatch of dialogue say) and image stand in a complementary relationship. The words, as well as the images, are fragments of a more general syntagm, and the unity of the message is realized at a higher level, namely, at the level of the story, the anecdote, the *diegesis*.[66]

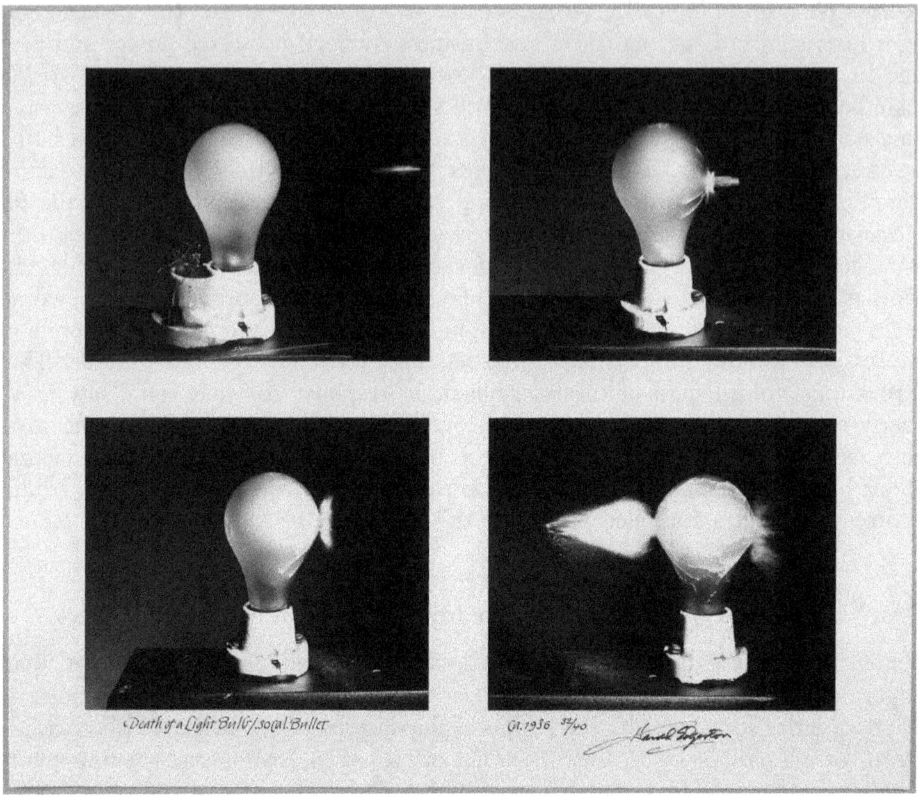

Figure 2.6 Harold Edgerton, *Death of a Light Bulb/.30 Caliber Bullet*, Print 32/40. 1936. (Image: 41.9 cm x 50.5 cm/16½ in. x 19⅞ in.) Harvard Art Museums/Fogg Museum, Gift of The Harold and Esther Edgerton Family Foundation, P1996.78.

In order to recreate the implied event of the light bulb's death – from the anticipation of being shot to its reality and aftermath – viewers project a narrative upon the set of photographs to achieve the *diegesis* Barthes describes.

While Edgerton's title implies the light bulb's "death," it is not obvious which image – if any – in the series reveals *the* final "death" moment. Perhaps the unseen, expected collapse of the bulb marks the "death." Maybe the bulb "dies" in the final photograph, as the bullet's damage has been done. It is unclear. In the first frame, a bullet that appears as a white blur approaches the light bulb from the left at more than twice the speed of sound.[67] The light bulb is unscathed. Time in the first frame of the series is represented as an anticipatory future tense, as viewers know that about 1/1,000,000th of a second later, the inanimate-but-personified light bulb will be struck by the projectile and "die."

As viewers approach Edgerton's *Death of a Light Bulb*, multiple relationships to time are conjured simultaneously, all of which have a personified death as a final consequence. André Bazin wrote that photography atomizes and "embalms time, rescuing it simply from its proper corruption."[68] The "corruption" to which Bazin refers is the violence of lived time's steadily flowing, relentless stream – which inevitably brings death to all of us. Photography halts time, sidestepping its cruel agency and the certainty of death. Barthes also suggested that all photographs remind viewers of the finality of death as an endpoint of the human experience of time. Barthes, for example, wrote that the photograph has a peculiar capacity to represent the past in the present, and thus to testify to the passing of time.[69] Therefore, Barthes argued, all photographs speak of our own death's inevitability in the future. Our fascination with the medium, Barthes suggests, is motivated by our own time anxiety, and by the fear of death.

Death of a Light Bulb similarly conjures more complex representations of photographic time. In the first image, viewers anticipate the bullet's contact with the light bulb, a meeting that is stilled in the photograph just before it occurred, but which transpired in 1936. In other words, we are looking at the *future* "death" of a light bulb that actually transpired *seventy-four years ago*. These images also function as icons of technology – specifically, of analog stroboscopy, a method that provided new visual access to the picosecond. The series continues to encapsulate the wonder inherent in that moment of photographic innovation for an audience in the present digital age. Tense distinction thus collapses upon itself.

Compression shock waves moving at 15,000 feet per second precede the bullet in the second image, making small cracks in the bulb as the bullet begins to enter the light bulb. A puff of glass dust and several cracks precede the bullet. The compression waves move faster than the projectile itself, lending the image a sense of past tense (*"the bullet lags behind the shock waves"*), present-ness (*"the bullet just entered the light bulb"*), and future anticipatory tense (*as viewers, in a Barthesian sense, await the full collapse and "death" of the light bulb*). In addition, the shock waves and bullet operate separately, yet simultaneously and in tandem, within the same microsecond-long exposure-time duration.

In Edgerton's third image, no bullet appears at all. The projectile is contained and momentarily suspended inside the light bulb, and viewers anticipate its continued *future* travel through the other side – and the completion of the light bulb's destruction and "death." The other side of the light bulb has begun to crack even before the bullet penetrates it, as a result of the compression wave of force that precedes the bullet. It foreshadows another certain *future* event, the collapse of the bulb. The trail of the bullet is marked by the small billowing of crushed glass powder, which trails the bullet passage and implies the *past*-tense event of the bullet's penetration of the bulb. But viewers also appreciate the peculiar tense stasis of the *present-tense* exposure-time duration, during which the bullet presumably is suspended in mid-flight inside of the bulb.

In the final image, the bullet has finished its travel through the light bulb, leaving the bulb's surface fully marred with cracks. The bullet's path is filled with an arrowhead-shaped residue of glass dust. A *past* tense (*marked by the physical proof of the bullet's travel through the light bulb*) is implied in this image, along with a *present* tense (*the bulb has not yet collapsed, but it is enjoying a fraction of a second of pre-collapse stasis*), and an anticipatory *future* tense (*the light bulb is going to crumble*). But Edgerton ends this series with this image, leaving further *future*-tense possibilities unspecified. Viewers are left to wonder: *What does the light bulb's collapse look like? Why does he deny us one more image? Where does the bullet go next?* Edgerton leaves these aspects of *Death of a Light Bulb* unresolved, denying his observers temporal closure. As such, a lingering uncertain future tense looms beyond the frames of this image.

Closer examination of the light bulbs in this series of photographs suggest that these images were not taken in consecutive microseconds, or with an even interval between exposures.[70] The first bulb (upper left panel) has a black wattage stamp on its top edge. This emblem is white in the second bulb (upper right panel). In addition, the base of the second bulb does not hold a second bulb, as the holder for the first bulb does. Although the third bulb (bottom left) has a white wattage stamp, its holder is near the left edge of the table. No table edges are visible in the second bulb's panel. The third bulb's base has been tiled slightly counter-clockwise, and the white nick on the table is farther from the base in the third panel than in the others. In addition, the trajectory of cracks made by the bullet's impact in panel two extend to the center of the bulb. In the third panel, the cracks stay confined to the right edges of the bulb. Continued entry of the bullet in panel three likely would have produced additional, not fewer, cracks. The bulb in the final (bottom right) panel is situated on a stand on the other side of the table nick from that which appears in the third and fourth panels. The black marking on the bulb holder from the other three panels is missing. These small variations suggest that the sequence is not sequential. Edgerton's alteration of the series reveals innate failings in the design of the stroboscope. Film could not be advanced in less than a microsecond, so no more than one exposure can occur for each light-bulb-shooting event.

Death of a Light Bulb therefore is comprised of four images of four *different* light bulbs as they are penetrated by four *different* bullets, from a session of photographing during which, most likely, dozens to hundreds of bulbs "died" in order for Edgerton to have enough images from which to compose a series. As an altered series, Edgerton's four chosen photographs reveal microseconds of non-consecutive, non-linear, creatively reconfigured time to make a new narrative: the *Death of a Light Bulb* (not of *four* light bulbs). The individual photographs may be (unverifiably) regarded as truthful to the microsecond-long appearance of spatial relations of a light bulb and bullet in the context of four separate light bulbs' "deaths," but the series is *not* truthful to the appearance of a *singular* bullet's passage through *a* light bulb. What, then, does the series of photographs offer, if not an objective document? This image depicts not what we *see*, but what we *know – bullets move through space with great speed* – even if we can only see the start and end of that movement with our unaided eyes. The photograph has an estranged relationship to perceived reality, and thus is an abstraction to our "lived"-time experience. Still, though, Edgerton's photographs are truthful to the appearance of a bullet in mid-flight as it approaches, enters, occupies, and leaves these four light bulbs.

Instead, then, of readily projecting a linear temporal narrative of imagined events that transpired between the frames to support the plot suggested by Edgerton's title (this is *a* light bulb's "death"), viewers may begin to imagine the additional narrative possibilities implied by the process by which the images were made. Calibrating the strobe lights with the firing of the gun doubtlessly consumed a significant amount of time for Edgerton. He needed time for the consideration of logistical concerns, and the creation of safeguards, such as creating a landing place for the bullets that would stop them before they could penetrate walls and endanger lives. He had to secure the bulbs' holders firmly to the table, set up the camera, and calibrate the discharge strobe lighting in synchronicity with the bullet's passage. Odds of catching the microsecond in which sequential steps of a bulb's "death" in four exposures were slim to none. What we see is assuredly the result of sustained work and much trial and error.

Editing the photographs from contact sheets to visually convey fairly even temporal intervals of the sequential steps of the "death" of *a* light bulb also was a time-consuming part of creating the series. Perhaps Edgerton was drawn by the first image's uncanny suspension of a bullet in mid-air, in mid-flight. Conceivably, the destructive, advancing force of the compression shock waves – inaccessible to the unaided human eye – likely inspired him to pick the second photograph. Capturing a visual representation of the decisive microsecond in which the bullet was contained entirely within the bulb must have involved technical prowess, incessant experimentation, and luck. Similarly, the final image represents a fragile, fully cracked light bulb on the verge of collapse. Perhaps the challenge of achieving such images motivated Edgerton to select a representation of these microseconds as the third and fourth photographs of the series. Knowledge of the altered nature of this series of photographs draws viewers' attention to Edgerton's hand in the image-making process. This, in turn, adds additional dimensions of time, as observers consider the temporal specificities of Edgerton's presence and various facets of his working processes. Close consideration of the *Death of a Light Bulb* four-photograph set thus reveals that, in addition to presenting estranged "realities" that we take on faith (for they are unverifiable by the unaided human eye), even microsecond-exposure stroboscopic photographs express time in many more ways than as a view of a transportable slice of the temporal past.

Photography, Instantaneity and Multifarious Temporalities

This chapter began with a discussion of Tiphaigne de la Roche's book *Giphantie*, an eighteenth-century science-fiction novel describing a "proto-photographic" image might have to say about how the medium expresses time, with respect to the temporality created by the visual cultures of painting and printmaking. In de la Roche's novel, a science-fictional "canvas" covered in reactive chemical ooze stills a convincingly "real" image of a storm in thirty minutes – an impossible turnaround for making a naturalistic painting. *Giphantie*'s description of the "elementary spirits'" stilling of an image in a "twinkle of an eye" reminds us of the radical departure of photography's expression of time from that of the medium of painting. It also begins, fundamentally, to recognize the medium's unusual, "atomized" views of time – which may be captured from a full field of space in a fraction of a moment, as many of the medium's earliest innovators attested. As they continued to experiment, they noted that the world changed as they made their stilled images. Sub-events unfolded, such as the movement of the Sun, or the movement of subjects in – and sometimes out of – the frame. Sub-events and differences between the known world and the photographed one disrupt the pervasive naturalism of the photograph, providing a tension that the image encourages and never resolves.

Indeed, for the first seventy years of its existence as a medium, photographers were preoccupied with pushing the very boundaries of photographic instantaneity. Exposure times steadily diminished in a matter of four decades, thanks to more reactive photochemistry, lenses, and shutter design. This endeavor culminated from

1870 to 1900 in the fraction-of-a-second motion studies of figures such as Muybridge, whose batteries of cameras captured 1/1,000th of a second. Edgerton, of course, would reduce this exposure to 1/1,000,000th of a second, using strobe lighting and no shutters.

However, to think of a photograph *only* as a frozen slice of time from the past, one that only allows the past to express the conditions of its achievement, would be limiting. As closer examination of Edgerton's *Death of a Light Bulb* suggested, even stroboscopic photographs – apexes of the achievement of instantaneous-time images – the temporal experience cannot be described as simply being from the past. Time, as it appears in photographs, is variegated and heterogenous, and has consequences beyond that which appears in the frame, or which can be described solely in terms of materialized exposure and process-related times.

In this sense, de la Roche's novel's title, *Giphantie*, is itself prescient. It suggests the book's theme: articulating a multifarious temporal experience that has three interpenetrating parts: "*what has past*," "*what is now passing*," and "*what will pass*." Our very experience with time is itself complex, consisting of an interplay of multiple temporalities. While photography suggests that its purview of time is objective, physical, material, and "actual," it is not a "real" view of time as we experience it in psychological/"lived" time. The time in photography is always an "other" to ours. And, as objects that may touch off narratives, they offer a plethora of additional temporal possibilities, as the next chapter will explore.

Notes

1 Charles-François Tiphaigne de la Roche, *Giphantie: Or, a View of What Has Passed, What is Now Passing, and During the Present Century, What Will Pass in the World* (London: Robert Horsfield, 1761), 93–100. This book originally was published in French one year earlier by the Paris publisher A Babylone. *Giphantie*, incidentally, is a word formed by the reordering of the letters of the author's name, "Tiphaigne."

2 Mayer and Pierson (Ernest Mayer, Frédéric Mayer and Pierre-Louis Pierson), *La Photographie: considérée comme art et comme industrie* (Paris: Librairie L. Hachette et Cie, 1862); Josef Maria Eder, *History of Photography* (New York: Dover Publications, Inc., 1945), 89–90; Georges Pontonniée, *The History of the Discovery of Photography* (New York: Arno Press, 1973), 42–46; Beaumont Newhall, ed., *Photography: Essays and Images* (New York: Museum of Modern Art, 1980), 13–14.

3 Tanya Sheehan and Andres Zervigon, *Photography and Its Origins* (New York: Routledge, 2014), 218; Geoffrey Batchen, *Burning With Desire: The Conception of Photography* (Cambridge, MA: MIT Press, 1997), 31–32, 225; Robert Hirsch, *Seizing the Light: A History of Photography* (New York: McGraw-Hill, 2000), 7; Naomi Rosenblum, *A World History of Photography*, 4th ed. (New York: Abbeville Press, 2007), 193.

4 The full inscription, as it originally appears in French: "L'Heliographie. Les premiers resultants obtenus Spontanément par l'action de la lumiere. Par Monsieur Nièpce, De Chalon sur Sâone. 1827."

5 Geoffrey Batchen, "Desiring Production," in *Each Wild Idea: Writing, Photography, History* (Cambridge, MA and London: MIT Press, 2001), 3–24, 192–196; Jennifer Green-Lewis, *Already the Past: Victorian Photography, Literature, and the Invention of Modern Memory* (New York and London: Bloomsbury Academic Press, 2017).

6 Batchen, "Desiring Production," 13–14, 22. John House also suggests that nineteenth-century artists' interest in time was multifarious. He notes that Constable's paintings, as well as French Realist and Naturalist paintings of peasants, make reference to agricultural cycles (cyclical time), the Impressionists' paintings of railways and train stations implicitly reference technological rapidity and change (instantaneity), and paintings such as William

Holman Hunt's *Awakening Conscience* (1853) convey a psychological/transformative time. John House, "Seasons and Moments: Time and Nineteenth-Century Art," in *The Story of Time*, Kristen Lippincott, ed. (London: Merrell Holberton, 1999), 194–197.

7 Batchen, "Desiring Production," 16–17, 23. As Peter Watson points out, the Romantic writers were keenly aware of the effects of industrial and technological modernity: "It is the younger romantics – Shelley, Keats and Byron – who adopt an uncompromising humanism, aware of the dehumanizing effects of factory life on life in general, and even more conservative representatives, Wordsworth and Scott, share their 'democratic' sympathies in that their work is aimed at the popularization – even the politicalization – of literature." Peter Watson, *Ideas: A History of Thought and Invention, From Fire to Freud* (New York: Harper Perennial, 2005), 612.

8 Batchen, "Desiring Production," 99.

9 Francis Bauer, Manuscript notations and labels on the verso of Joseph Nicéphore Niépce's *View From the Window at Le Gras*, c. 1827–1828.

10 Daguerre is quoted in: Helmut Gernsheim, *The Origins of Photography* (New York: Thames and Hudson, 1982), 38.

11 For Daguerre's description of the making of a diorama, please see: Louis Jacques Mandé Daguerre, *An Historical and Descriptive Account of the Various Processes of the Daguerreotype and the Diorama* (New York: Winter House Ltd., 1971), 81–86.

12 Daguerre, *An Historical and Descriptive Account of the Various Processes of the Daguerreotype and the Diorama*, 81–86.

13 Sophie Thomas, "Making Visible: The Diorama, the Double and the (Gothic) Subject," *Romantic Circles Praxis Series* (2005): n.p.

14 The Inflation Calculator, www.westegg.com/inflation/infl.cgi. Accessed June 20, 2018.

15 Hippolyte Gaucheraud, "The Fine Arts: A New Discovery" (1839), in Newhall, ed., *Photography: Essays and Images*, 17–18.

16 Beaumont Newhall, ed., *Photography: Essays and Images*, 17–18. This quote is originally was published in: *La Gazette de (Paris) France*, January 6, 1839.

17 Daguerre's photographic experiments were made public by the French government on August 19, 1839. An instruction manual in French, English, Italian, Spanish, and Swedish was immediately printed and distributed: Louis Jacques Mandé Daguerre, *History and Process of Photogenic Drawing by Means of the Daguerreotype* (London, 1839), 7. For an English version of this manual, please see: Louis Jacques Mandé Daguerre, *On a New Process of Polishing Plates Destined to Receive Photographic Images Permitting Identical Results to Be Obtained So Long as the External Circumstances Remain the Same* (London, 1843). The distribution of these manuals assured the spread of the daguerreotype process, and further irritated William Henry Fox Talbot and Hippolyte Bayard – who were among a cluster of individual inventors who claimed to have invented a photographic process before 1839. For more about the competing claims of the simultaneous inventors of photography – and the problems with anointing one man as *the* inventor of the medium, please see: Batchen, *Burning With Desire*, 24–53, 223–231.

18 Samuel F. B. Morse, "The Daguerrotipe," *New York Observer*, April 20, 1839, 62. The spelling of the title of Morse's article has not been changed. For Daguerre, achieving the shortest possible exposure, in which less movement and blur still would be discernible, was not just a matter of achieving realism (which was a concern of his), but also to make the process more economically viable. Daguerre noted that in a letter he sent to Niépce's son Isidore in 1834: "At the point where we are now, we must agree that reproduction is still limited because it still takes three or four hours. It would not be the same if we had a more immediacy and success would not be doubtful.'" Keith James Laidler, *To Light Such a Candle: Chapters in the History of Science and Technology* (Oxford: Oxford University Press, 1998), 11. Laidler cites this original source: François Arago, "Le daguerreotype," in CRAS, session of August, 19, 1839, t. 9, (París: Bachelier, 1835–1965): 250–267, 256.

19 Gaucheraud is quoted in: Helmut and Alison Gernsheim, *L. J. M. Daguerre: The History of the Diorama and the Daguerreotype* (New York: Dover Publications, 1968), 85. The original article appeared in: Hippolyte Gaucheraud. *Gazette de France*, January 6, 1839. This is the Gernsheims' translation.

20 Although this English-language publication would have reached Talbot, it is more likely that he learned of Daguerre's invention from Gaucheraud's January 12, 1839, report in the *Literary Gazette*: Roger Taylor, *Impressed by Light: British Photographs From Paper Negatives, 1840–1860* (New York: Metropolitan Museum of Art, 2007), 397, fn. 5. Talbot claimed to have started his experimentation with photography in 1834 in this essay: William Henry Fox Talbot, *Some Account of the Art of Photogenic Drawing* (London: R. and J. E. Taylor, 1839).

21 William Henry Fox Talbot, *The Pencil of Nature*, Part I (London, 1844), n.p.

22 Talbot, *The Pencil of Nature*, n.p.

23 William Henry Fox Talbot, "Some Account of the Art of Photographic Drawing, Or, the Process by Which Natural Objects May Be Made to Delineate Themselves Without the Aid of the Artist's Pencil" (1839), in *Photography: Essays and Images*, Beaumont Newhall, ed. (New York: Museum of Modern Art, 1980), 25. As Talbot notes, in his earliest experiments: "An hour's exposure on a summer afternoon left only the impress of the highlights on the paper." Trying other lenses allowed him to decrease exposure times. He comments on his work with "mouse trap" cameras (1835): "After the lapse of half an hour, I gathered them all up, and brought them within doors to open them. When opened, there was in each a miniature picture of the objects before which it had been placed." Beaumont Newhall, *The History of Photography, From 1839 to the Present*, 5th ed. (New York: Museum of Modern Art, 1982), 20. Newhall quotes William Henry Fox Talbot, "Some Account of the Art of Photographic Drawing" (1839), in Newhall, *Photography: Essays and Images*, 23–30.

24 Michael Sand, ed. *Specimens and Marvels: William Henry Fox Talbot and the Invention of Photography* (New York: Aperture Foundation, 2000), 36.

25 Gaucheraud, in Helmut and Alison Gernsheim, *L. J. M. Daguerre*, 85.

26 Newhall, *The History of Photography*, 124; Similar times appear in this chart: Helmut Gernsheim, *The Origins of Photography*, 582.

27 Newhall, *The History of Photography*, 124.

28 Newhall, *The History of Photography*, 124, 303; Marc Oute, "Gelatine," *British Journal Photographic Almanac* (1881): 213.

29 Newhall, *The History of Photography*, 19.

30 Newhall, *The History of Photography*, 117–124.

31 Newhall, *The History of Photography*, 119.

32 Newhall, *The History of Photography*, 119–120. Photographer/writer Jerome O'Neill argues, based on his own technical experiments, that Muybridge's exposure times may have been embellished: "Muybridge claimed shutter speeds as high as 1/5,000 second with wet collodion plates, but sufficient exposure would have been a problem. In his later work he found that 1/600 to 1/800 second was adequate to 'stop' the stride of a horse, and of course photo finish pictures are concerned primarily with the horse's nose, which is moving at a slower rate than the legs." Jerome P. O'Neill, "The History of Race Photo Finish Photography: Interim Report of the Photo Finish Research Project" (Rochester: 1972), 13.

33 Newhall, *The History of Photography*, 128. Newhall mentions further developments in the camera that improved its ability to conquer the momentary, including the 1925 introduction of the glass flash bulb, which was developed by Paul Vierkötter in 1925: With this innovation, photography was no longer constrained by the quality of available light. Newhall, *The History of Photography*, 231. In this text, the stroboscopic/flash-enabled work of Harold Edgerton and Barbara Morgan is discussed as a result of flash developments, and without reference to Talbot's earlier Leyden-jar experiments: Newhall, *The History of Photography*, 232–233.

34 Newhall, *The History of Photography*, 129; Michel Frizot, *A New History of Photography* (Köln: Könemann, 1998), 238.

35 Miles Orvell, *The Real Thing: Imitation and Authenticity in American Culture, 1880–1940* (Chapel Hill: University of North Carolina Press, 1989), 100.

36 Charles Baudelaire, "The Painter of Modern Life" (1863), in *Baudelaire: Selected Writings on Art and Artists*, ed. P. E. Charvet (Cambridge: Cambridge University Press, 1981), 402. In the essay "Salon de 1859," Baudelaire praised photography's ability to aid the

sciences and the arts, but only as "their very humble servant." He warned against viewing photography as the ideal fulfillment of the goals to which painting aspired: Charles Baudelaire, "Salon de 1859: Le Public Moderne et la Photographie (1859)" in *Oeuvres complètes* (Paris: Gallimard, 1961), 2.

37 Marta Braun, *Picturing Time: The Work of Étienne-Jules Marey (1830–1904)* (Chicago: University of Chicago Press, 1992). Philip Prodger, François Dagognet, and Rebecca Solnit have offered similar studies of Muybridge: Phillip Prodger, *Time Stands Still: Muybridge and the Instantaneous Photography Movement* (Oxford: Oxford University Press, 2003); Rebecca Solnit, *River of Shadows: Eadweard Muybridge and the Technological Wild West* (New York: Penguin Books, 2004); François Dagognet, *Étienne-Jules Marey: A Passion for the Trace* (Cambridge, MA: MIT Press, 1992).

38 O'Neill, "The History of Race Photo Finish Photography: Interim Report of the Photo Finish Research Project." [unpublished manuscript] Rochester, NY: 1972, 10.

39 Max Perchick, "Extending the Frontiers of Photographic Vision," *PSA Journal* 62, No. 11 (1996): 26.

40 Solnit, 81.

41 Solnit, 11, 81.

42 Guy Debord, *The Society of the Spectacle*, Donald Nicholson-Smith, trans. (New York: Zone Books, 1994), 92.

43 Gail Buckland, *Fox Talbot and the Invention of Photography* (London: Scholar Press, 1980), 107–118, 202; Chitra Ramalingam, "Stopping Time: Henry Fox Talbot and the Origins of Freeze-Frame Photography," *Endeavour* 32, No. 3 (September 2008): 86.

44 For more about Talbot's experiments with "instantaneous" photography, please see: Larry Schaaf, "Sparks, Spinning Wheels, and Whites of Egg," *The Talbot Catalogue Raisonné*, posted June 17, 2016. Available at: https://talbot.bodleian.ox.ac.uk/2016/06/17/sparks-spinning-wheels-whites-of-egg/

45 Gail Buckland, *Fox Talbot and the Invention of Photography*, 107–118.

46 Schaaf, "Sparks, Spinning Wheels, and Whites of Egg," n.p.

47 Schaaf, "Sparks, Spinning Wheels, and Whites of Egg," n.p.

48 Gail Buckland, *Fox Talbot and the Invention of Photography*, 220; Charles Vernon Boys, *Electric-Spark Photographs of Flying Bullets* (Washington, DC: Smithsonian Institution), 1893.

49 Gail Buckland, *Fox Talbot and the Invention of Photography*, 220; Arthur Mason Worthington, *A Study of Splashes* (London: Longmans, Green, and Co, 1908).

50 Harold Edgerton and James R. Killian, Jr., *Moments of Vision: The Stroboscopic Revolution in Photography* (Cambridge, MA: MIT Press, 1979), 6.

51 Harold Edgerton and K. J. Germeshausen, "Stroboscopic Photography," *Electronics* 5, No. 1 (July 1932): 220–221; Harold Edgerton and K. J. Germeshausen, "The Mercury Arc as an Actinic Stroboscopic Light Source," *Review of Scientific Instruments* 3, No. 10 (October 1932): 535–542; Edgerton and Killian, *Moments of Vision*, 2. James R. Killian, Jr., who co-authored the text *Moments of Vision: The Stroboscopic Revolution in Photography* with Edgerton, was the editor of *Technology Review* at the time Edgerton's photographs were published in that magazine.

52 Harold Edgerton and K. J. Germeshausen, "Catching the Click With the Stroboscope," *American Golfer*, November 1933.

53 "Interesting People," *American Magazine*, May 1934, 48.

54 "Portrait," *Fortune*, November 1936.

55 "Speaking of Pictures ... This Is a Cock Fight," *Life*, December 14, 1936, 2–3; "Speaking of Pictures ... Stroboscopic Lights Make Action Stand Still," *Life*, November 20, 1939, 10–13; "Speaking of Pictures ... These Show Science True and Beautiful," *Life*, July 28, 1941, 9–11; "Speaking of Pictures ... These Culminate 70 Years of High-Speed Photography," *Life*, October 27, 1941, 13–15.

56 "Speaking of Pictures ... Stroboscopic Lights Make Action Stand Still," 10–13. Gjon Mili met Edgerton while the two were graduate students at the Massachusetts Institute of Technology. Mili had regularly contributed photographs as a freelancer to *Life* since the magazine's founding in 1936, and he aided Edgerton in making the right connections to get his work published in the magazine. Edgerton and Killian, *Moments of Vision*, 2.

57 For more about Edgerton's appeal to the masses, see: Marta Braun, "Time and Photography," in *Tempus Fugit: Time Flies*, Jan Schall, ed. (Kansas City, MO: Nelson-Atkins Museum of Art, 2000), 132–143, 143.

58 Harold Edgerton and James R. Killian, Jr., *Flash! Seeing the Unseen by Ultra High Speed Photography* (Boston: Hale, Cushman & Flint, 1939).

59 Edgerton remained in the public eye throughout his career. In 1940, Edgerton collaborated on the film *Quicker Than a Wink* with Pete Smith. It won an Academy Award. He also directed the use of the strobes for nighttime aerial reconnaissance photography during World War II, and received a Medal of Freedom in 1946 for his work. Edgerton developed Rapatronic cameras for the photographing of nuclear explosions in the following year, and worked over several decades to develop various sonar devices for photographing the ocean floor. One of those devices was responsible for locating the sunken *Titanic* off the coast of Nova Scotia in 1986.

60 Angelika Schedel, "An Unprecedented Sensation – Public Reaction to the Discovery of X-Rays," *Physics Education*, 30, No. 6 (Nov. 1995): 343. Probably the most popular novel about X-rays was H. G. Wells's *The Invisible Man* (1897), which was based on the notion that a man's exposure to X-rays could make his body invisible: H. G. Wells, *The Invisible Man* (1897) (New York: New American Library/Penguin Books, 2002).

61 Edgerton and Killian, *Moments of Vision*, 1.

62 Braun, "Time and Photography," 136.

63 Mary Ann Doane, *The Emergence of Cinematic Time: Modernity, Contingency, the Archive* (Cambridge, MA and London: Harvard University Press, 2002), 4.

64 Frederick W. Taylor, *The Principles of Scientific Management* (New York: 1911), 36.

65 Doane, *The Emergence of Cinematic Time*, 6.

66 Roland Barthes, "Rhetoric of the Image," *Image, Music, Text*. Stephen Heath, ed. and trans. (New York: Hill and Wang, 1977), 41.

67 For technical details about the making of this image, please see: Stephen Dalton, *Split Second: The World of High-Speed Photography* (Salem: Salem House, 1984), 26–27.

68 André Bazin, "The Ontology of the Photographic Image," *Film Quarterly*, 13, No. 4 (Summer, 1960): 4.

69 Roland Barthes, *Camera Lucida: Reflections on Photography*, Richard Howard, trans. (New York: Hill and Wang, 1981), 96.

70 This possibility was first raised by this author in: Kris Belden-Adams, "Harold Edgerton and Complications of the 'Photographic Instant,'" *Frame*, Spring 2011: http://framejournal.org/view-article/14. Since the publication of that essay, M.I.T.'s Edgerton archives have added this to their entry for the panels: "Though the series appears to be of a single bullet smashing a single light bulb, it is actually a composite of four separate bullets and bulbs." https://webmuseum.mit.edu/detail.php?module=objects&type=related&kv=96455

Bibliography

Ades, Dawn. "Art and Time in the Twentieth Century." In *The Story of Time*. London: Merrell Holberton, 1999. 202–211.

Aldur, Martin. "The Representation of Movement in Art." *Burlington Magazine for Connoisseurs*, Vol. 23, No. 124 (1913): 204–207.

Allen, Richard. "Representation, Illusion and the Cinema." *Cinema Journal*, Vol. 32, No. 2 (1993): 21–48.

Antliff, Robert Mark. "Bergson and Cubism: A Reassessment." *Art Journal*, Vol. 47, No. 4 (1988): 341–349.

Antliff, Robert Mark. "The Fourth Dimension and Futurism: A Politicized Space." *The Art Bulletin*, Vol. 82, No. 4 (2000): 720–733.

Arnheim, Rudolf. "The Two Authenticities of the Photographic Media." *Journal of Aesthetics and Art Criticism*, Vol. 51, No. 4 (1993): 537–540.

Arnold, H. J. P. *William Henry Fox Talbot: Pioneer of Photography and Man of Science*. London: Hutchinson Bentham, 1977.

Baer, Ulrich. "The Shutter's Suspension of Time: Toward a Democritean Theory of Photography." In *Time Expanded.* Madrid: La Fábrica/PHotoEspaña, 2010. 43–51.

Baetens, Jan and van de Gelder, Hilde, eds. *Photography and Time.* Leuven: Leuven University Press, 2010.

Baker, George. "Photography's Expanded Field." *October*, Vol. 114 (2005): 120–140.

Barthes, Roland. *Mythologies.* New York: Hill and Wang, 1972.

Barthes, Roland. *Image, Music, Text.* New York: Hill and Wang, 1977.

Barthes, Roland. *Camera Lucida: Reflections on Photography.* New York: Hill and Wang, 1981.

Batchen, Geoffrey. *Burning with Desire: The Conception of Photography.* Cambridge, MA: MIT Press, 1997.

Barthes, Roland. *Each Wild Idea: Writing, Photography, History.* Cambridge: MIT Press, 2001.

Barthes, Roland. "Electricity Made Visible." In *New Media, Old Media: A History and Theory Reader*, Wendy Hui Kyong Chun and Thomas W. Keenan, eds. New York: Routledge, 2005. 27–44.

Barthes, Roland. *Forget Me Not: Photography and Remembrance.* New York: Princeton Architectural Press, 2006.

Barthes, Roland. *Photography: Degree Zero: Reflections on Roland Barthes's Camera Lucida.* Cambridge, MA: MIT Press, 2009.

Batchen, Geoffrey and Heiferman, Marvin, eds. *Now Is Then: Snapshots from the Maresca Collection.* New York: Princeton Architectural Press, 2007.

Baudelaire, Charles. "Salon de 1859: Le Public Moderne et la Photographie (1859)." In *Oeuvres Completes, Vol. 2.* Paris: Gallimard, 1961. 222–224.

Baudelaire, Charles. *The Painter of Modern Life and Other Essays*, Jonathan Mayne, trans. New York: Da Capo, 1964.

Baudrillard, Jean. "The Implosion of Meaning in the Media and the Implosion of the Social in the Masses." In *The Myths of Information*, Kathleen Woodward, ed. Madison: Coda Press, 1980.

Baudrillard, Jean. *Simulations.* New York: Semiotext(e), 1983.

Bauman, Zygmunt. *Liquid Modernity.* Cambridge, MA: Polity Press, 2000.

Bazin, André. *What Is Cinema?.* Berkeley, Los Angeles and London: University of California Press, 1967.

Bazin, André. "The Ontology of the Photographic Image" (1967). In *Classic Essays on Photography*, Alan Trachtenberg, ed. New Haven: Leete's Island Books, 1980. 237–244.

Belden-Adams, Kris. "Harold Edgerton and Complications of the 'Photographic Instant,'" *Frame* (Spring 2011), http://framejournal.org/view-article/14.

Bellour, Raymond. *Le temps d'un mouvement: Aventures et mesaventures de l'instant photographique.* Paris: Centre National de la Photographie, 1986.

Bellour, Raymond. "The Film Stilled." *Camera Obscura*, Vol. 8, No. 3 (24) (1990): 98–123.

Belting, Hans. "Image, Culture, History, Media, Visualization, Art." *Critical Inquiry*, Vol. 31, No. 2 (2005): 302–319.

Benjamin, Walter. "Baudelaire". In *Selected Writings, Vol. 1, 1913–1926*, Marcus Bullock and Michael W. Jennings, eds. Rodney Livingstone, trans. Cambridge and London: The Belknap Press of Harvard University, 1922. 361–362.

Benjamin, Walter. "Little History of Photography". In *Selected Writings, Vol. 2, 1927-1934*, Edmund Jephcott and Kingsley Shorter, trans. Cambridge and London: The Belknap Press of Harvard University, 1931. 507–530.

Benjamin, Walter. "The Work of Art in an Age of Its Technological Reproducibility" (1936–1939). In *Selected Writings, Vol. 4, 1938–1940*, Harry Zohn and Edmund Jephcott, trans. Cambridge and London: The Belknap Press of Harvard University, 2003. 251–283.

Benjamin, Walter. *Illuminations.* Hannah Arendt, ed. New York: Schocken, 1969.

Benjamin, Walter. *The Arcades Project*, Howard Eiland and Kevin LcLaughlin, trans. Cambridge: Harvard University Press, 1999a.

Bergson, Henri. *Creative Evolution*. Arthur Mitchell, trans. New York: Henry Holt & Co., 1911.

Bergson, Henri. "Form and Becoming." In *The Philosophy of Time: A Collection of Essays*, Richard M. Gale, ed. New Jersey: Humanities Press, 1911. 397–405.

Bergson, Henri. "Time and Free Will: An Essay on the Immediate Data of Consciousness." In *Gilles Deleuze and the Ruin of Representation*, Dorothea E. Olkowski, ed. Berkeley: University of California Press, 1999a. 121.

Borradori, Giovanna. "The Temporalization of Difference: Reflections on Deleuze's Interpretation of Bergson." *Continental Philosophy Review*, Vol. 34 (2001): 1–20.

Braun, Marta. *Picturing Time: The Work of Étienne-Jules Marey (1830-1904)*. Chicago and London: University of Chicago Press, 1992.

Braun, Marta "The Expanded Present: Photographing the Moment." In *Beauty of Another Order: Photography in Science*, Ann Thomas, ed. New Haven: Yale University Press, 1997. 150–185.

Braun, Marta. "Time and Photography." In *Tempus Fugit: Time Flies*, Jan Schall, ed. Kansas City, MO: Nelson-Atkins Museum of Art, 2000. 132–143.

Brubaker, David. "Andre Bazin on Automatically Made Images." *The Journal of Aesthetics and Art Criticism*, Vol. 51, No. 1 (1993): 59–67.

Bruce, Roger, ed. *Seeing the Unseen: Dr. Harold Edgerton and the Wonders of Strobe Alley*. Rochester, NY: The Publishing Trust of George Eastman House, 1994.

Buckland, Gail. *First Photographs: People, Places and Phenomena as Captured for the First Time by the Camera*. New York: Macmillan Publishing, 1980.

Buckland, Gail. *Fox Talbot and the Invention of Photography*. London: Scholar Press, 1980.

Burgin, Victor. "The Eclipse of Time." In *Photography and Time*, Jan Baetens and Hilde van de Gelder, eds. Leuven: Leuven University Press, 2010. 125–140.

Cadava, Eduardo. "The Image: A Monster of Time." In *Time Expanded*. Madrid: La Fábrica/ PHotoEspaña, 2010. 27–41.

Canales, Jimena. *A Tenth of a Second: A History*. Chicago: University of Chicago Press, 2009.

Cardwell, Sarah. "About Time: Theorizing Adaptation, Temporality and Tense." *Literature/ Film Quarterly*, Vol. 31, No. 2 (2003): 82–92.

Carey, James W. *Communication as Culture: Essays on Media and Society*. Boston: Unwin Hyman, 1989.

Carr, Herbert Wildon. "'Time' and 'History' in Contemporary Philosophy; With Special Reference to Bergson and Croce." *Proceedings of the British Academy* 8 (1921): 331–349.

Cazort, Mimi. "Photography's Illustrative Ancestors: The Printed Image." In *Beauty of Another Order: Photography in Science*, Ann Thomas, ed. New Haven: Yale University Press, 1997. 14–25.

Crary, Jonathan. *Techniques of the Observer: On Vision and Modernity in the Nineteenth Century*. Cambridge, MA: MIT Press, 1990.

Crary, Jonathan. *Suspensions of Perception: Attention, Spectacle, and Modern Culture*. Cambridge, MA: MIT Press, 1999.

Curie, Gregory. "Visible Traces: Documentary and the Contents of Photographs." *Journal of Aesthetics and Art Criticism*, Vol. 57, No. 3 (1999): 285–297.

Dagognet, François. *Étienne-Jules Marey: A Passion for the Trace*. New York: Zone Books, 1992.

Daguerre, Louis Jacques Mandé. *History and Process of Photogenic Drawing by Means of the Daguerreotype*. London: W. Strange, 1839.

Daguerre, Louis Jacques Mandé. *On A New Process of Polishing Plates Destined to Receive Photographic Images Permitting Identical Results to be Obtained So Long as the External Circumstances Remain the Same*. London: R. Hastings, 1843.

Daguerre, Louis Jacques Mandé. *An Historical and Descriptive Account of the Various Processes of the Daguerreotype and the Diorama*. New York: Winter House Ltd., 1971.
Dalton, Stephen. *Split Second: The World of High-Speed Photography*. Salem: Salem House, 1984.
Darius, Jon. *Beyond Vision*. Oxford: Oxford University Press, 1984.
De Duve, Thierry. "Time Exposure and Snapshot: The Photograph as Paradox." *October* Vol. 5 (1978): 113–125.
De la Roche, Charles-Tiphaigne, Françoise. *Giphantie: Or, A View of What Has Passed, What Is Now Passing, and During the Present Century, What Will Pass in the World*. London: Robert Horsfield, 1761.
De Man, Paul. "Roland Barthes and the Limits of Structuralism." *Yale French Studies*, Vol. 77 (1990): 177–190.
Debord, Guy. *The Society of the Spectacle*, Donald Nicholson-Smith, trans. New York: Zone Books, 1994.
Deleuze, Gilles. "Mediators." In *Incorporations*, Jonathan Crary and Sanford Kwinter, eds. New York: Zone, 1992. 281–294.
Delpire, Robert. "Avant-propos: Tout ou rien." In *Le temps d'un mouvement: Aventures et mesaventures de l'instant photographique*, Raymond Bellour, ed. Paris: Centre National de la Photographie, 1986.
Derrida, Jacques. "Signature, Event, Context." *Glyph* (1): 1977, 180.
Derrida, Jacques. *Of Grammatology (1976)*. Baltimore: Johns Hopkins University Press, 1981.
Derrida, Jacques. "No Apocalypse, Not Now (Full Speed Ahead, Seven Missiles, Seven Missives)." *Diacritics*, Vol. 14, No. 2 (1984), 20–31.
Derrida, Jacques. *Mémoires: For Paul de Man Reading*. New York: Columbia University Press, 1989.
Derrida, Jacques. "Right of Inspection." *Art & Text*, Vol. 32 (1989): 20–97.
Derrida, Jacques. *Copy, Archive, Signature: A Conversation on Photography*. Stanford, CA: Stanford University Press, 2010.
Doane, Mary Ann. "Information, Crisis, Catastrophe." In *The Logics of Television: Essays in Cultural Criticism*, Patricia Mellencamp, ed. Bloomington: Indiana University Press, 1990.
Doane, Mary Ann. "Temporality, Storage, Legibility: Freud, Marey and the Cinema." *Critical Inquiry*, Vol. 22, No. 2 (Winter 1996): 313–343.
Doane, Mary Ann. *The Emergence of Cinematic Time: Modernity, Contingency, the Archive*. Cambridge, MA and London: Harvard University Press, 2002.
Doane, Mary Ann. "Real Time: Instantaneity and the Photographic Imaginary." In *Stillness and Time: Photography and the Moving Image*, David Lowry and Joanna Green, eds. Brighton: Photoworks/Photoforum, 2006. 22–38.
Druckey, Timothy, ed. *Electronic Culture: Technology and Visual Representation*. New York: Aperture Foundation, 1996.
Duttlinger, Carolin. *Kafka and Photography*. Oxford: Oxford University Press, 2007.
Duttlinger, Carolin. "Imaginary Encounters: Walter Benjamin and the Aura of Photography." *Poetics Today*, Vol. 29, No. 1 (Spring 2008), 79–101.
Eder, Josef. *History of Photography*, Edward Epstean, trans. New York: Dover Publications Inc., 1972.
Edgerton, Harold. *Stopping Time: The Photographs of Harold Edgerton*. New York: Harry N. Abrams, Inc., 1987.
Edgerton, Harold and Germeshausen, K. J. "Stroboscopic Photography." *Electronics*, Vol. 5, No. 1 (July 1932): 220–221.
Edgerton, Harold and Germeshausen, K. J. "The Mercury Arc as an Actinic Stroboscopic Light Source." *Review of Scientific Instruments*, Vol. 3, No. 10 (October 1932): 535–542.
Edgerton, Harold and Germeshausen, K. J. "Catching the Click With the Stroboscope." *American Golfer* (November 1933), 17.

Edgerton, Harold. "Speaking of Pictures . . . Stroboscopic Lights Make Action Stand Still." *Life* (November 20 1939): 10–13.

Edgerton, Harold. "Speaking of Pictures . . . These Culminate 70 Years of High-Speed Photography." *Life* (October 27, 1941): 13–15.

Edgerton, Harold and Killian, James R., Jr. *Flash! Seeing the Unseen by Ultra High Speed Photography*. Boston: Hale, Cushman & Flint, 1939.

Edgerton, Harold and Killian, James R., Jr. *Moments of Vision: The Stroboscopic Revolution in Photography*. Cambridge, MA: MIT Press, 1979.

Elias, Norbert. *Time: An Essay*, Edmund Jephcott, trans. Cambridge: Blackwell, 1994.

Elkins, James, ed. *Photography Theory*. New York: Routledge, 2007.

Ellenbogen, Josh. *Photography and the Imperceptible: Bertillon, Galton, Marey*. Ph.D. diss., University of Chicago, 2005.

Ellenbogen, Josh. "Authority, Objectivity, Evidence: Scientific Photography in Victorian Britain." *Studies in the History and Philosophy of Science*, Part C Vol. 39, No. 1 (Part C 2008): 171–175.

Ellenbogen, Josh. "Camera and Mind." *Representations*, Vol. 101 (Winter, 2008): 86–115.

Ellul, Jacques. *The Technological Society*, trans. John Wilkinson. New York: Knopf, 1964.

Faulkner, Keith W. *Deleuze and the Three Syntheses of Time*. New York: Peter Lang, 2006.

Faye, Jan, Scheffler, Uwe, and Urchs, Max, eds. *Perspectives on Time*. Norwell, MA: Kluwer Academic Publishers, 1997.

Friday, Jonathan. "Photography and the Representation of Vision." *Journal of Aesthetics and Art Criticism*, Vol. 59, No. 4 (2001): 351–362.

Frizot, Michel. "Un instant, s'il vous plaît . . ." In *Le temps d'un mouvement: Aventures et mesaventures de l'instant photographique*. Paris: Centre National de la Photographie, 1986. 7–13.

Frizot, Michel. "Speed of Photography: Movement and Duration." In *A New History of Photography*, Koln, Germany: Konemann, 1998. 242–257.

Frizot, Michel. "A Critical Discussion of the Historiography of Photography." *Arken Bulletin*, Vol. 1 (2002): 58–65.

Galassi, Peter. *Before Photography: Painting and the Invention of Photography*. New York: Museum of Modern Art, 1981.

Gehlen, Arnold. *Man in the Age of Technology*. New York: Columbia University Press, 1980.

Gere, Charlie. *Art, Time and Technology*. Oxford and New York: Berg, 2006.

Gernsheim, Helmut. *The Origins of Photography*. New York: Thames and Hudson, 1982.

Gernsheim, Helmut and Gernsheim, Alison. *L. J. M. Daguerre: The History of the Diorama and the Daguerreotype*. New York: Dover Publications, 1968.

Goldberg, Vicki. *The Power of Photography: How Photography Changed Our Lives*. New York: Abbeville Publishing Group, 1991.

Gombrich, Ernst Hans Josef. "Moment and Movement in Art." *Journal of the Warburg and Courtauld Institutes*, Vol. 27 (1964): 293–306.

Green, David. "Marking Time: Photography, Film and Temporalities of the Image." In *Stillness and Time: Photography and the Moving Image*. Brighton: Photoworks/Photoforum, 2006. 9–21.

Green, David and Lowry, Joanna. *Stillness and Time: Photography and the Moving Image*. Brighton: Photoworks/Photoforum, 2006.

Green-Lewis, Jennifer. *Already the Past: Victorian Photography, Literature, and the Invention of Modern Memory*. New York and London: Bloomsbury Academic Press, 2017.

Griffiths, Alison. "Wondrous Difference: Cinema, Anthropology, and Turn-of-the-Century Visual Culture." In *Film and Culture*, John Belton, ed. New York: Columbia University Press, 2002.

Grosz, Elizabeth, ed. *Becomings: Explorations in Time, Memory, and Futures*. Ithaca and London: Cornell University Press, 1999.

Guattari, Félix. "Regimes, Pathways, Subjects." In *Incorporations*, Jonathan Crary and Sanford Kwinter, eds. New York: Zone, 1992. 16–37.

Gubser, Michael. "Time and History in Alois Riegl's Theory of Perception." *Journal of the History of Ideas*, Vol. 66, No. 3 (2005): 451–474.

Guerlac, Suzanne. "Thinking in Time: Henri Bergson." *MLN*, Vol. 120, No. 5 (2005): 1091–1098.

Guido, Laurent and Lugon, Oliver, eds. *Between Still and Moving Images*. Bloomington: Indiana University Press, 2012.

Gunning, Tom. "From Kaleidoscope to the X-ray: Urban Spectatorship, Poe, Benjamin and Traffic in Souls." *Wide Angle*, Vol. 19, No. 4 (1913): 25–63.

Gunning, Tom. "The 'Arrested' Instant: Between Stillness and Motion." In *Between Still and Moving Images*, Laurent Guido and Oliver Lugon, eds. Bloomington: Indiana University Press, 2012. 23–31.

Gutierrez, Tania Ruiz. *Morphoscopy of the Transient*, Department of Art & Sciences of Art. Université Paris I, Pantheon-Sorbonne, 2001.

Harbison, Martha. "A Brief, Recent Flash in the History of Photography." *Popular Science*, Vol. 263, No. 2 (2003), 54.

Hardison, O. B. *Disappearing Through the Skylight: Culture and Technology in the Twentieth Century*. New York: Viking Penguin, 1989.

Heidegger, Martin. *Being and Time*. Joan Stambaugh, trans. Albany: State University of New York Press, 1927.

Heidegger, Martin. "The Age of the World Culture." In *Electronic Culture*, Timothy Druckey, ed. New York: Aperture Foundation, 1996. 46–61.

Henderson, Linda Dalrymple. *The Fourth Dimension and Non-Euclidean Geometry in Modern Art*. Princeton: Princeton University Press, 1983.

Hendricks, Gordon. *Eadweard Muybridge: The Father of the Motion Picture*. New York: Grossman Publishers, 1975.

Herzog, Amy. "Images of Thought and Acts of Creation: Deleuze, Bergson, and the Question of Cinema." *InVisible Culture*, No. 3 (2000): 1–19.

Himmelfarb, Gertrude. *The Roads to Modernity: The British, French and American Enlightenments*. New York: Alfred A. Knopf, 2004.

House, John. "Seasons and Moments: Time and Nineteenth-Century Art." In *The Story of Time*, Kristen Lippincott, ed. London: Merrell Holberton, 1999. 194–197.

Humphreys, Kathryn. "Looking Backward: History, Nostalgia, and American Photography." *American Literary History*, Vol. 5, No. 4 (1993): 686–699.

Husserl, Edmund. *The Phenomenology of Internal Time-Consciousness*. Martin Heidegger, ed. Bloomington and London: Indiana University Press, 1966.

Husserl, Edmund. *The Crisis of the European Sciences and Transcendental Phenomenology: An Introduction to Phenomenological Philosophy*. David Carr, trans. Evanston: Northwestern University Press, 1970.

Huyssen, Andreas. *After the Great Divide: Modernism, Mass Culture, Postmodernism*. Bloomington and Indianapolis: Indiana University Press, 1986.

Huyssen, Andreas. *Twilight Memories: Marking Time in a Culture of Amnesia*. London and New York: Routledge, 1994.

Huyssen, Andreas. "Present Pasts: Media, Politics, Amnesia." *Public Culture*, Vol. 12, No. 1 (2000): 21–38.

Ierley, Merritt. *Wondrous Contrivances: Technology at the Threshold*. New York: Clarkson Potter Publishers, 2002.

Ippolito, Jon. "Trusting Aesthetics to Prosthetics." *Art Journal*, Vol. 56, No. 3 (1997): 68–74.

Izenberg, Gerald N. *Impossible Individuality: Romanticism, Revolution and the Origins of Modern Selfhood, 1787-1802*. Princeton: Princeton University Press, 1992.

Jameson, Fredric. "The End of Temporality." *Critical Inquiry*, Vol. 29 (2003): 695–718.

Jenkins, Reese. *Images and Enterprises: Technology and the American Photographic Industry, 1839 to 1925*. Baltimore: Johns Hopkins University Press, 1987.
Jeu de Paume. *L'Ombre du temps*. Paris: Ministère de la Culture et de la Communication, 2004.
Johnson, Geraldine. A. *Sculpture and Photography: Envisioning the Third Dimension*. Cambridge (UK) and New York: Cambridge University Press, 1998.
Johnson, Richard. "Historical Returns: Transdisciplinary, Cultural Studies, History." *European Journal of Cultural Studies*, Vol. 4, No. 3 (2001): 261–288.
Johnston, Adrian. *Time Driven: Metapsychology and the Splitting of the Drive*. Evanston, IL: Northwestern University Press, 2005.
Jonsson, Bodil. *Unwinding the Clock: Ten Thoughts on Our Relationship to Time*. New York: Harcourt, Inc., 1999.
Keck, Frederic. "The Virtual, the Symbolic, and the Actual in Bergsonian Philosophy and Durkheimian Sociology." *MLN*, Vol. 120, No. 5 (2005): 1133–1145.
Kern, Stephen. *The Culture of Time and Space, 1880-1918*. Cambridge: Harvard University Press, 1983.
Kern, Stephen. "Time and Art in Twentieth-Century Culture." In *Tempus Fugit: Time Flies*, Jan Schall, ed. Kansas City, MO: Nelson-Atkins Museum of Art, 2000c. 22–34.
Knight, Diana. *Barthes and Utopia: Space, Travel, Writing*. Oxford: Clarendon, 1997.
Kracauer, Siegfried. "Photography." *Critical Inquiry*, Vol. 19 (1993), 421–436.
Kracauer, Siegfried. *The Mass Ornament: Weimar Essays*. Cambridge, Mass.: Harvard University Press, 1995.
Krauss, Rosalind. "The Im/Pulse to See." In *Vision and Visuality*, Hal Foster, ed. Seattle: Bay Press, 1988. 51–75.
Krauss, Rosalind. *The Optical Unconscious*. Cambridge: MIT Press, 1993.
Krauss, Rosalind. *The Originality of the Avant-Garde and Other Modernist Myths*. Cambridge: MIT Press, 1999.
Krauss, Rosalind. "Reinventing the Medium." *Critical Inquiry*, Vol. 25, No. 2 (1999): 289–305.
Krauss, Rosalind. *A Voyage on the North Sea: Art in the Age of the Post-Medium Condition*. New York: Thames and Hudson, 2000.
Kriebel, Sabine T. "Theories of Photography: A Short History." In *Photography Theory*, James Elkins, ed. New York: Routledge, 2007. 17–18.
Kuipers, Dean and Aitken, Doug. *I Am A Bullet: Scenes from an Accelerating Culture*. New York: Crown Publishers, 2000.
Kulvicki, John. "Image Structure." *The Journal of Aesthetics and Art Criticism*, Vol. 61, No. 4 (2003): 323–340.
Laidler, Keith James. *To Light Such a Candle: Chapters in the History of Science and Technology*. Oxford and New York: Oxford University Press, 1998.
Landes, David S. *Revolution in Time: Clocks and the Making of the Modern World*. Cambridge, MA and London: Harvard University Press, 1983.
Lawrence, Amy. "Counterfeit Motion: The Animated Films of Eadweard Muybridge." *Film Quarterly*, Vol. 57, No. 2 (2004): 15–25.
Lee, Pamela M. "Some Kinds of Duration: The Temporality of Drawing as a Process Art." In *Afterimage: Drawing through Process*, Cornelia H. Butler, ed. Cambridge, Mass.: MIT Press, 1999. 25–48.
Lee, Pamela M. *Chronophobia: On Time and Art of the 1960s*. Cambridge, MA, and London: MIT Press, 2004.
Lienhard, John. *Inventing Modern: Growing Up with X-rays, Skyscrapers, and Tailfins*. New York: Oxford University Press, 2003.
Lippincott, Kristen. *The Story of Time*. London: Merrell Holberton, 1999.
Lovejoy, Margot. "Art, Technology and Postmodernism: Paradigms, Parallels and Paradoxes." *Art Journal*, Vol. 49, No. 3 (1990): 257–265.

Lunenfeld, Peter, ed. *The Digital Dialectic: New Essays on New Media*. Cambridge: MIT Press, 1999.

MacDonnell, Kevin. *Eadweard Muybridge: The Man Who Invented the Moving Picture*. Boston and Toronto: Little, Brown and Company, 1972.

Mandel, Mike. *Making Good Time: Scientific Management - the Gilbreths, Photography and Motion, Futurism*. Riverside, Calif.: Museum of Photography, The University of California, Riverside, 1989.

Marvin, Carolyn. *When Old Technologies Were New: Thinking About Electric Communication in the Late Nineteenth Century*. New York: Oxford University Press, 1988.

Marx, Leo. *The Machine in the Garden*. New York: Oxford University Press, 1964.

Mayer, Louis Frédéric and Pierson, Pierre Louis. *La Photographie: Considérée comme art et comme industrie*. Paris: Librairie L. Hachette et Cie., 1862.

Maynard, Patrick. "Talbot's Technologies: Photographic Depiction, Detection and Reproduction." *The Journal of Aesthetics and Art Criticism*, Vol. 47, No. 3 (1989): 263–276.

Maynard, Patrick. "Photographic Seeing." In *The Engine of Visualization: Thinking through Photography*. Ithaca, NY and London: Cornell University Press, 1997. 191–227.

McCauley, Elizabeth Anne. *Industrial Madness: Commercial Photography in Paris, 1848-1871*. New Haven: Yale University Press, 1994.

McElhone, John P. "The Signature of Light: Photo-Sensitive Materials in the Nineteenth Century." In *Beauty of Another Order: Photography in Science*, Ann Thomas, ed. New Haven: Yale University Press, 1997. 60–75.

McQuire, Scott. *Visions of Modernity: Representation, Memory, Time and Space in the Age of the Camera*. London: Sage Publications, 1998.

Merloo, Joost. *Along the Fourth Dimension; Man's Sense of Time and History*. New York: John Day Co., 1970.

Mitchell, William J. *The Reconfigured Eye: Visual Truth in the Postphotographic Era*. Cambridge: MIT Press, 1992.

Morris, Edward, ed. *Constable's Clouds: Paintings and Cloud Studies by John Constable*. Edinburgh and Liverpool: National Galleries of Scotland and the National Museums and Galleries on Merseyside, 2000.

Morse, Samuel F. B. "The Daguerrotipe." *New York Observer*, Vol. 17, April 20, 1839, 62.

Muybridge, Eadweard. "Improvement in the Method and Apparatus for Photographing Objects in Motion," United States Patent Office. United States of America: 1879.

Muybridge, Eadweard. *Animal Locomotion: The Muybridge Work at the University of Pennsylvania*. Philadelphia: J. B. Lippincott Co., 1888.

Muybridge, Eadweard. *Animals in Motion*. London: Chapman and Hall, 1899.

Newhall, Beaumont. "Photography and the Development of Kinetic Visualization." *Journal of the Warburg and Courtauld Institutes*, Vol. 7 (1944): 40–45.

Newhall, Beaumont. *An Historical and Descriptive Account of the Various Processes of the Daguerreotype and the Diorama*. New York: Winter House Ltd., 1971.

Newhall, Beaumont, ed., *Photography: Essays and Images*. New York: The Museum of Modern Art, 1980.

Newhall, Beaumont. *The History of Photography, From 1839 to the Present*, fifth edition. New York: Museum of Modern Art, 1982.

Olkowski, Dorothea E. "Time Lost, Instantaneity and the Image." *Parallax: A Journal of Metadiscursive Theory and Cultural Practices*, Vol. 9, No. 1 (2003): 28–38.

Orvell, Miles. "Reproduction and 'The Real Thing': The Anxiety of Realism in the Age of Photography." In *The Technological Imagination: Theories and Fictions*, Teresa De Lauretis, Andreas Huyssen, and Kathleen Woodward, eds. Madison: Coda Press Inc., 1980. 49–64.

Orvell, Miles. *The Real Thing: Imitation and Authenticity in American Culture, 1880-1940*. Chapel Hill: University of North Carolina Press, 1989.

Osborne, Peter. "Photography in an Expanding Field: Distributive Unity and Dominant Form." In *Where is the Photograph?* Brighton: Photoforum and Photoworks, 2003. 47–62.

Ott, John. *My Ivory Cellar: The Story of Time-Lapse Photography*. Chicago: Twentieth Century Press, Inc., 1958.

Oute, Marc. "Gelatine." *British Journal Photographic Almanac* (1881): 213.

Peirce, Charles Sanders. *Collected Papers of Charles Sanders Peirce*, Charles Hartstone and Paul Weiss, eds. Cambridge: Harvard University Press, 1932.

Peirce, Charles Sanders. "Logic as Semiotic: The Theory of Signs." In *Semiotics: An Introductory Anthology*, Robert E. Innis, ed. Bloomington: Indiana University Press, 1985. 1–23.

Pontonniée, Georges. *The History of the Discovery of Photography*. New York: Arno Press, 1973.

Rabinbach, Anson. *The Human Motor: Energy, Fatigue, and the Origins of Modernity*. Berkeley: University of California Press, 1990.

Rabinbach, Anson. "Neurasthenia and Modernity." In *Incorporations*, Jonathan Crary and Sanford Kwinter, eds. New York: Zone, 1992. 178–189.

Ramalingam, Chitra. "Stopping Time: Henry Fox Talbot and the Origins of Freeze-Frame Photography." *Endeavour*, Vol. 32, No. 3 (September 2008): 86.

Ratliff, Floyd. "Illusions in Man and His Instruments." *Journal of Philosophy*, Vol. 68, No. 19 (1971): 591–597.

Rawson, Philip. *Art and Time*. Cranbury, NJ: Associated University Press, 2005.

Ray, Christopher. *Time, Space and Philosophy*. New York and London: Routledge, 1991.

Ribiere, Mireille and Baetens, Jan, eds. *Time, Narration, and the Fixed Image*. Amsterdam and Atlanta: Rodopi, 2001.

Robins, Kevin. "The Virtual Unconscious in Photography." In *Electronic Culture*, Timothy Druckey, ed. New York: Aperture Foundation, 1996. 154–163.

Sand, Michael, ed. *Specimens and Marvels: William Henry Fox Talbot and the Invention of Photography*. New York: Aperture Foundation, 2000.

Sartre, Jean-Paul. *L'imaginaire: Psychologie-Phinoménologique de L'imagination*. Paris, 1940; Bernard Frechtman, trans., *The Psychology of Imagination*. New York: Philosophical Library, 1948.

Schaaf, Larry J. *Out of the Shadows: Herschel, Talbot and the Invention of Photography*. New Haven and London: Yale University Press, 1992.

Schaaf, Larry J. *The Correspondence of William Henry Fox Talbot: A Draft Calendar*. Glasgow: Glasgow University Library, 1995.

Schaaf, Larry J. *Records of the Dawn of Photography: Talbot's Notebooks P & Q*. Cambridge: Cambridge University Press, 1996.

Schaaf, Larry J. "Invention and Discovery." In *Beauty of Another Order: Photography in Science*, Ann Thomas, ed. New Haven: Yale University Press, 1997. 26–59.

Schaaf, Larry J. "Sparks, Spinning Wheels, and Whites of Egg," *The Talbot Catalogue Raisonné*, Posted June 17, 2016. Available at: https://talbot.bodleian.ox.ac.uk/2016/06/17/sparks-spinning-wheels-whites-of-egg/

Schall, Jan, ed. *Tempus Fugit: Time Flies*. Kansas City, Mo.: Nelson-Atkins Museum of Art, 2000.

Scharf, Aaron. "Painting, Photography, and the Image of Movement." *Burlington Magazine for Connoisseurs*, Vol. 104, No. 710 (May 1962): 186, 188–195.

Schedel, Angelika. "An Unprecedented Sensation – Public Reaction to the Discovery of X-Rays." *Physics Education*, Vol. 30, No. 6 (November. 1995): 343.

Schwartz, Sanford. *The Matrix of Modernism: Pound, Elliot, and Early Twentieth-Century Thought*. Princeton: Princeton University Press, 1985.

Sconce, Jeffrey. *Haunted Media: Electronic Presence from Telegraphy to Television*. Durham: Duke University Press, 2000.

Sheehan, Tanya and Zervigon, Andres. *Photography and Its Origins*. New York: Routledge, 2014.

Sherover, Charles M. and Johnson, Gregory. *'Are We In Time?' and Other Essays on Time and Temporality*. Evanston, Ill.: Northwestern University Press, 2003.

Silverman, Kenneth. *Lightning Man: The Accursed Life of Samuel B. Morse*. New York: Alfed A. Knopf, 2003.

Simmel, Georg. "The Metropolis and Mental Life." In *The Sociology of Georg Simmel*, Kurt H. Wolff, ed. London: Collier-Macmillan, 1950. 409–424.

Simonsen, Dorthe Gert. "Accelerating Modernity: Time-Space Compression in the Wake of the Aeroplane." *The Journal of Transport History*, Vol. 26, No. 2 (September 2005): 98–117.

Snow, C. P. *The Two Cultures and the Scientific Revolution*. New York and London: Cambridge University Press, 1993.

Solnit, Rebecca. *River of Shadows: Eadweard Muybridge and the Technological Wild West*. Middlesex, England: Viking Penguin, 2003.

Stiegler, Bernard. *Technics and Time, 1: The Fault of Epimetheus*. Stanford: Stanford University Press, 1998.

"Stopping Time." *Science News*, Vol. 133, No. 2 (January 9, 1988).

Sutton, Damien. *Photography, Cinema, Memory: The Crystal Image of Time*. Minneapolis: University of Minnesota Press, 2009.

Sutton, Damien. "Delay. Notes on Photography as Non-Representational Thinking." In *Time Expanded*. Madrid: La Fábrica/PHotoEspaña, 2010. 145–151.

Szarkowski, John. *The Photographer's Eye*. New York: Museum of Modern Art, 1966.

Tagg, John. *The Burden of Representation: Essays on Photographies and Histories*. Amherst: University of Massachusetts Press, 1988.

Talbot, William and Henry Fox. "Some Account of the Art of Photogenic Drawing, or, the Process By Which Natural Objects May Be Made to Delineate Themselves Without the Aid of the Artist's Pencil". In *Photography: Essays and Images*, Beaumont Newhall, ed. New York: The Museum of Modern Art, 1839. 23–31.

Talbot, William and Henry Fox. *The Pencil of Nature* (1844). New York: Da Capo Press, 1844.

Taylor, Frederick W. *The Principles of Scientific Management*. New York: Harper & Brothers, 1911.

Taylor, Roger. *Impressed by Light: British Photographs from Paper Negatives, 1840-1860*. New York: The Metropolitan Museum of Art, 2007.

Thomas, Alan. *The Expanding Eye: Photography and the Nineteenth-Century Mind*. London: Croom Helm, 1978.

Thomas, Sophie. "Making Visible: The Diorama, the Double and the (Gothic) Subject," *Romantic Circles Praxis Series*. 2005. Last modified December 2015. www.rc.umd.edu/praxis/gothic/thomas/thomas.html

Thompson, Edward Palmer. "Time, 'Work-Discipline, and Industrial Capitalism'." *Past and Present* No. 38 (December 1967): 56–97.

University of Pennsylvania. *Animal Locomotion: The Muybridge Work at the University of Pennsylvania*. Philadelphia: J. B. Lippincott Co., 1888.

Vargish, Thomas and Mook, Delo E. *Inside Modernism: Relativity Theory, Cubism, Narrative*. New Haven and London: Yale University Press, 1999.

Watson, Peter. *Ideas: A History of Thought and Invention, from Fire to Freud*. New York: Harper Perennial, 2005.

Wike, Lori. "Photographs and Signatures: Absence, Presence, and Temporality in Barthes and Derrida." *InVisible Culture*, No. 3 (2000): 1–28.

Worthington, Arthur Mason. *The Romance of Science: The Splash of a Drop*. London: Society for Promoting Christian Knowledge, 1895. Reprint of the report delivered to the Royal Institution of Great Britain on May 18, 1894.

3 The Fluidity of "Narrative Time"

The *Leighton Family* (Fig. 3.1a [front], b [back]) is typical of many group portraits made by studio photographers in the southern United States at the fin de siècle. In it, a father, Samuel Humphrey Leighton (seated, front row, second from left), mother, Cynthia Belle Finch Leighton (also seated, front row, second from right), and their eight children, Nancy Ellen, Laura Hazel, Samuel David, Jacob Lafayette, Elbye Ann, Grover Cleveland, Lilly Lucinda, and Isa Belle Leighton – pose in two loosely organized rows, and each subject meets the direct gaze of the camera.[1] Nobody smiles, although hints of a smirk appear on Isa Belle's face. (As Christina Kotchemidova has suggested, smiling would not become a standard family-portrait convention until World War I, after a marketing blitz by the Kodak Corporation encouraged subjects to smile in order to look "alive."[2]) The youngest Leighton children sit still and appear well-behaved, as portrait-photography customs then preferred.[3] Shortly after this photograph was taken, it would become fashionable for parents to photograph their brood in the act of play, in order to convey that a family had enough money to allow children to enjoy the luxury of leisure time.[4]

The Leightons, an Oklahoma-frontier-settling, farming/ranching family, are dressed in their "Sunday best." The boys and men wear high-buttoned suits, and the women wear crisply tailored formal dresses with high, modest, banded collars, squared shoulders, and long sleeves – a style that was fashionable among pioneers in the 1890s.[5] Two of the young women wear jewelry, and hints of decorative trim on their bodices and waistlines are visible. The youngest two children – one male, one female – are perched in their parents' laps in the front row, and wear matching polka-dot frocks. The Leightons' conscious choice to wear new formal clothing conveys the family's social status and class-consciousness.[6]

The unobtrusive painted backdrop of *Leighton Family* features a double classical column that is met by a vertical row of geometric tiles and the hint of palm-tree foliage. While these architectural details and the presence of a sprig of tropical plant growth lend exoticism to this staged group portrait of a Southern/Midwestern farming family, no written evidence suggests that the Leightons – who were considered well-traveled because they had journeyed to Iowa and Kansas on the Chisholm Trail and the Rock Island Railroad – had ever seen Greece, Italy or tropical plants first-hand by the time the photograph was made in the early months of 1899.[7]

The ten Leightons cluster together to fit in the picture plane while allowing each family member's face to be seen. They are physically a tight-knit, cohesive group as

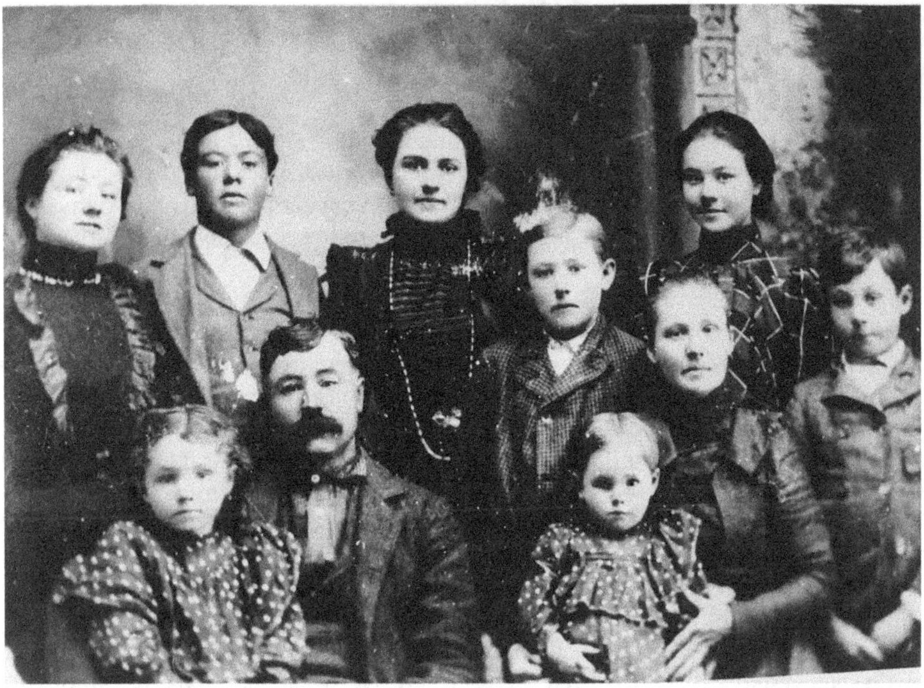

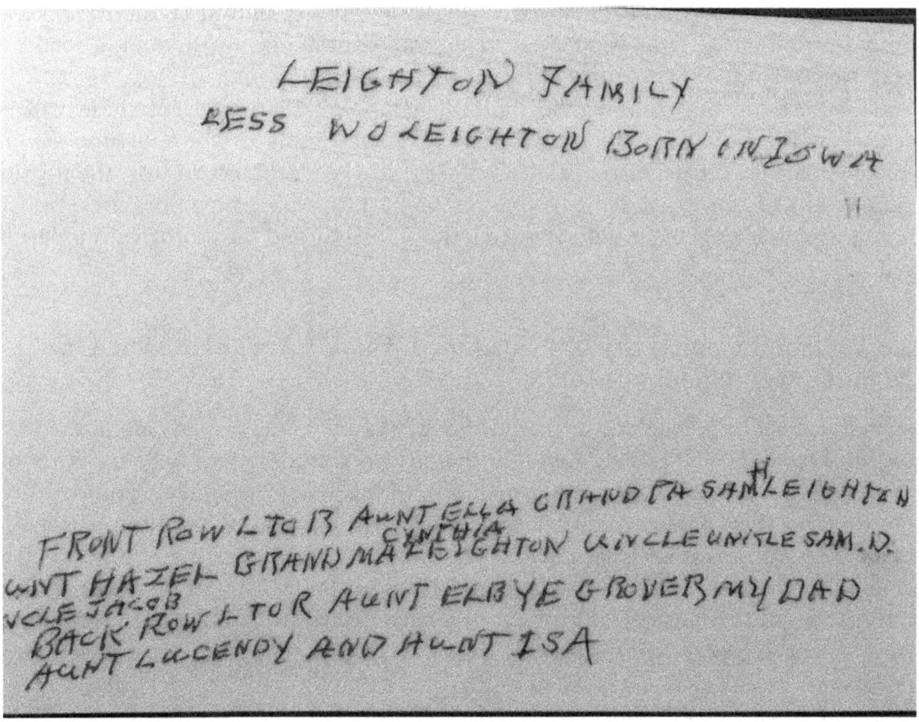

Figure 3.1a, 3.1b Photographer Unknown, *Leighton Family* (front and back), 1899. Notes added by Samuel Bernard Leighton. Courtesy of Karen Leighton Belden.

they present a staged, idealized rhetoric of family unity that had become a convention in studio group family portraiture. Barthes suggests that the act of self-presentation in photographs is yet more complex and riddled with questions of inauthenticity, because it consists of at least four considerations:

> In front of the lens, I am at the same time: *the one I think I am, the one I want others to think I am, the one the photographer thinks I am, and the one he makes use of to exhibit his art.* In other words, a strange action: each time I am (or let myself be) photographed, I invariably suffer from a sensation of inauthenticity, sometimes of imposture.[8]

Some of Barthes's considerations are linked to self-presentation and staging, some to the photographer's gaze, and others are linked to the afterlife of the photograph as an object that circulates to an audience of family, friends, and complete strangers – the photographer's potential customers, as well as later owners of the photograph. By lending themselves to this "strange action" of being photographed that Barthes describes, the Leightons succumb to the projection and dissemination of various "selves" that the image simultaneously rehearses and creates. But each of these "selves" enjoys a unique and changeable temporal specificity, and are not mutually exclusive. The photographic subject's considerations of who "I think I am," who "I want others to think I am," and "the one the photographer thinks I am" exist and evolve in a cognitively embedded sense of time as "temporality." In addition, as Barthes suggests, the *"selves"* that viewers perceive when the image is exhibited in various contexts are observer-dependent and enjoy interpretive temporal fluidity, a distinctive "narrative time." Moreover, viewers' placement of a family portrait within a non-linear, non-sequential temporal continuum by adding a fictional narrative convolutes distinctions between *time* and *temporality*.

In the sections that follow, the multiple temporal dimensions of these various "*selves*" will be discussed in relationship to a case study of the *Leighton Family* portrait. Barthes's categories of "selves" present a springboard for the further examination of the expanded possibilities of narrative-based temporalities, and of photography's engagement with fin-de-siècle industrial and technological modernity itself.

The Temporal Complexity of "The One I Think I Am" and "The One I Want Others to Think I Am"

Nancy Ellen ("Ella") Leighton Miller (1893–1988) sits in her father's lap in the front row (first from the left) of the *Leighton Family* portrait. She was only six years old at the time the picture was made, which would make her an unlikely source of the only surviving first-person narrative to speak for the day this photograph was made. Nonetheless, an unspecified number of years later, Miller wrote a poem titled "The Gumps's Pitcher Tooken" (Fig. 3.2a [page 1] and b [page 2]).[9] In it, she provides a rare first-person, hindsight-informed, narrative from an adult's perspective that places the taking of *Leighton Family* within a specific temporal continuum that is tinged with nostalgia and representations of the family as *"the one I think I am"* and as *"the one I want others to think I am."*

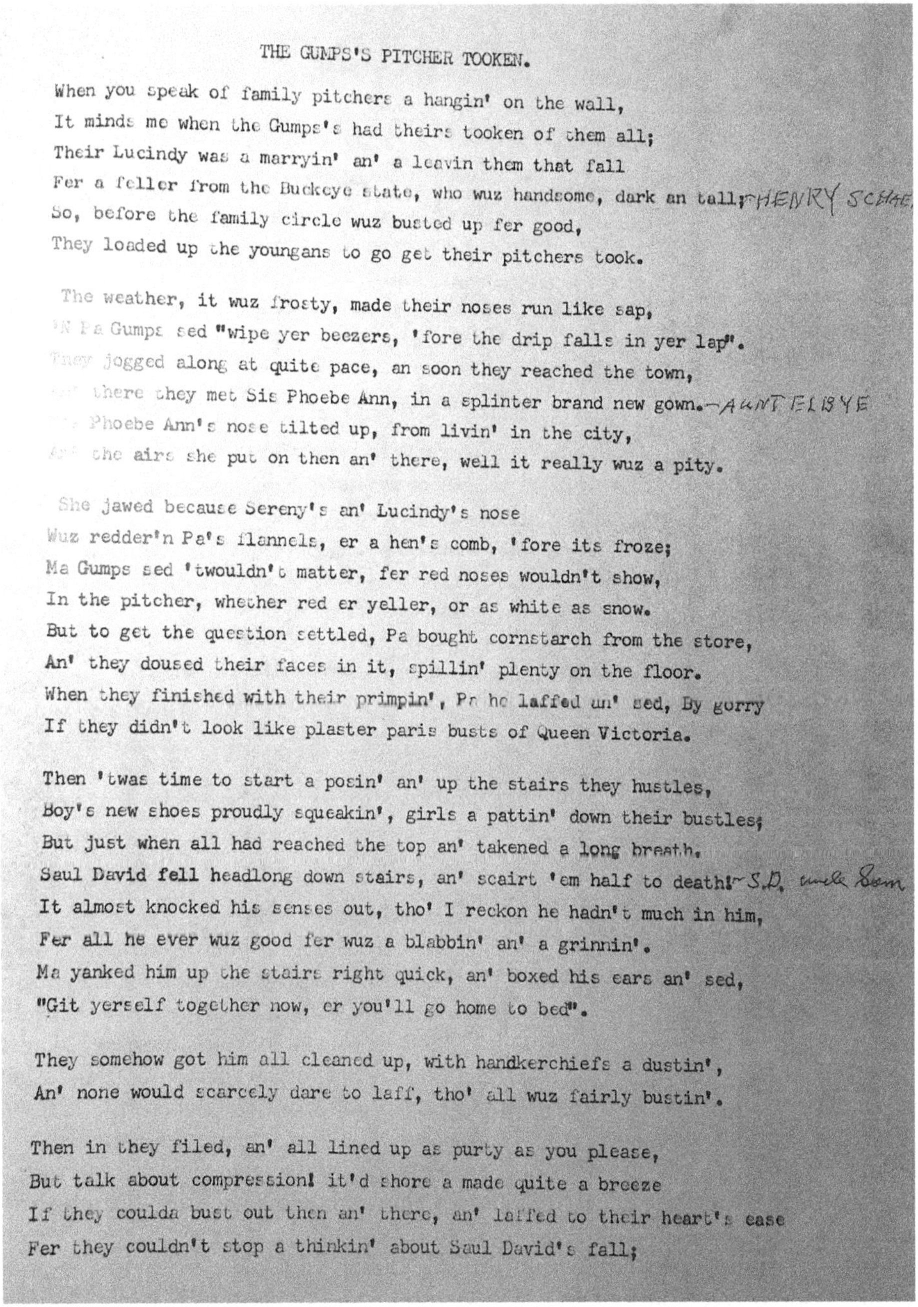

THE GUMPS'S PITCHER TOOKEN.

When you speak of family pitchers a hangin' on the wall,
It minds me when the Gumps's had theirs tooken of them all;
Their Lucindy was a marryin' an' a leavin them that fall
Fer a feller from the Buckeye state, who wuz handsome, dark an tall;—HENRY SCHAE
So, before the family circle wuz busted up fer good,
They loaded up the youngans to go get their pitchers took.

The weather, it wuz frosty, made their noses run like sap,
'N Pa Gumps sed "wipe yer beezers, 'fore the drip falls in yer lap".
They jogged along at quite pace, an soon they reached the town,
An' there they met Sis Phoebe Ann, in a splinter brand new gown.—AUNT ELBYE
Phoebe Ann's nose tilted up, from livin' in the city,
An' the airs she put on then an' there, well it really wuz a pity.

She jawed because Sereny's an' Lucindy's nose
Wuz redder'n Pa's flannels, er a hen's comb, 'fore its froze;
Ma Gumps sed 'twouldn't matter, fer red noses wouldn't show,
In the pitcher, whether red er yeller, or as white as snow.
But to get the question settled, Pa bought cornstarch from the store,
An' they doused their faces in it, spillin' plenty on the floor.
When they finished with their primpin', Pa he laffed an' sed, By gorry
If they didn't look like plaster paris busts of Queen Victoria.

Then 'twas time to start a posin' an' up the stairs they hustles,
Boy's new shoes proudly squeakin', girls a pattin' down their bustles;
But just when all had reached the top an' takened a long breath,
Saul David fell headlong down stairs, an' scairt 'em half to death!—S.D. uncle Sam
It almost knocked his senses out, tho' I reckon he hadn't much in him,
Fer all he ever wuz good fer wuz a blabbin' an' a grinnin'.
Ma yanked him up the stairs right quick, an' boxed his ears an' sed,
"Git yerself together now, er you'll go home to bed".

They somehow got him all cleaned up, with handkerchiefs a dustin',
An' none would scarcely dare to laff, tho' all wuz fairly bustin'.

Then in they filed, an' all lined up as purty as you please,
But talk about compression! it'd shore a made quite a breeze
If they coulda bust out then an' there, an' laffed to their heart's ease
Fer they couldn't stop a thinkin' about Saul David's fall;

Figure 3.2a, 3.2b Nancy Ellen Leighton Miller, "The Gumps's Pitcher Tooken" (Unpublished Poem, with Notes added by Samuel Bernard Leighton), n.d. Courtesy of Karen Leighton Belden.

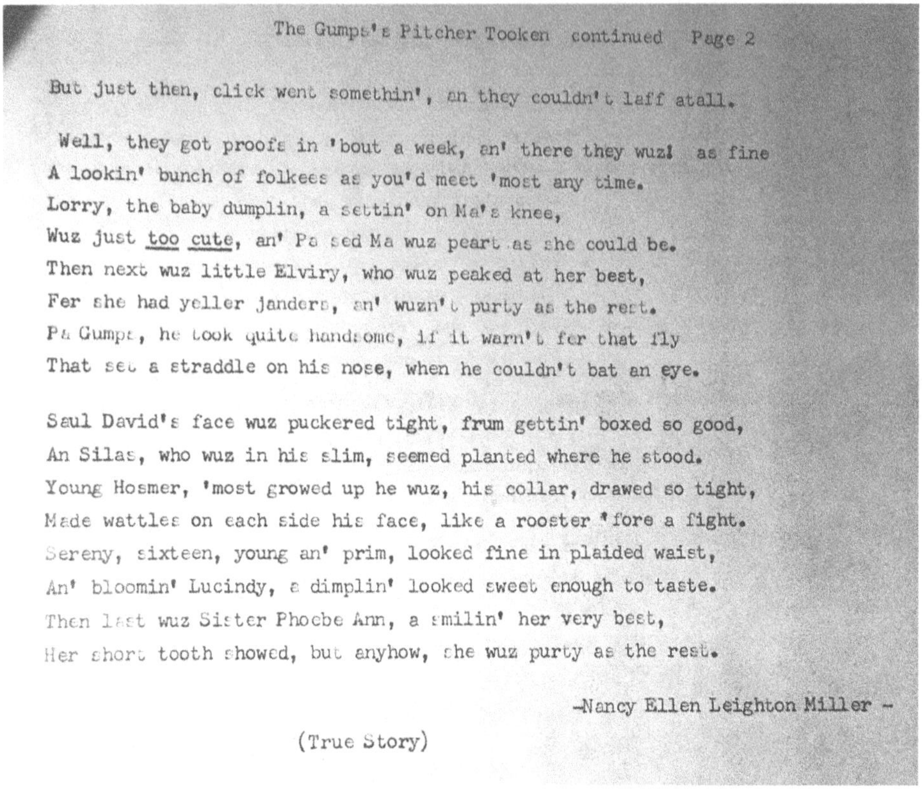

The Gumps's Pitcher Tooken continued Page 2

But just then, click went somethin', an they couldn't laff atall.

Well, they got proofs in 'bout a week, an' there they wuz! as fine
A lookin' bunch of folkees as you'd meet 'most any time.
Lorry, the baby dumplin, a settin' on Ma's knee,
Wuz just too cute, an' Pa sed Ma wuz peart as she could be.
Then next wuz little Elviry, who wuz peaked at her best,
Fer she had yeller janders, an' wuzn't purty as the rest.
Pa Gumps, he took quite handsome, if it warn't fer that fly
That set a straddle on his nose, when he couldn't bat an eye.

Saul David's face wuz puckered tight, frum gettin' boxed so good,
An Silas, who wuz in his slim, seemed planted where he stood.
Young Hosmer, 'most growed up he wuz, his collar, drawed so tight,
Made wattles on each side his face, like a rooster 'fore a fight.
Sereny, sixteen, young an' prim, looked fine in plaided waist,
An' bloomin' Lucindy, a dimplin' looked sweet enough to taste.
Then last wuz Sister Phoebe Ann, a smilin' her very best,
Her short tooth showed, but anyhow, she wuz purty as the rest.

-Nancy Ellen Leighton Miller -

(True Story)

Figure 3.2a, 3.2b (Continued)

Connecting to Grand Historical Narratives

"The Gumps's Pitcher Tooken" is the first of thirteen poems arranged in an unpublished booklet that has been handed down to descendants. In the poems, Miller wrote about topics as varied as Kansas's majestic and rugged landscape and its renegade politics, President Franklin Delano Roosevelt's Scottish terrier Falla, the tragic loss of lives during World War II, national patriotism, herself as a child, and her earliest memories of her youth on the Oklahoma frontier – where the *Leighton Family* photograph was made in 1899. As a whole, the poems nimbly navigate from nostalgic memories of family folklore, to references of broader historical narratives, such as Queen Victoria, the settlement of the Oklahoma frontier (in which the Leightons participated), and notorious Dalton Gang of outlaws (whose parents and siblings farmed about 10 miles from the Leightons). Personal narratives and grand narratives intermingle, and often are inseparable, suggesting that Miller viewed her family – and wished for it likewise to be seen in posterity – as a group whose activities were consequential to the history and settlement of the American frontier. Miller's humanizing narratives position the *Leighton Family* – as well as Miller herself – as "everyday people" who nonetheless were important players in the settlement of the United States frontier. Photography historian Liz Wells argues that family photographs can prompt individuals to re-enact the plights of ancestors, recapture personal history, and therefore provide a context for individuals' lives.[10] This effectively collapses distinctions between Barthes's categories of *"the one I think I am"* and *"the one I want others to think I am."* In this sense alone, the *Leighton Family* has multiple temporal implications, all inextricably tied to narratives implied by the photograph and provided by viewers in its ongoing reception.

According to family folklore and genealogy records, the first Leightons to arrive in America, Thomas and Joanna, came from Glasgow, Scotland, in 1633, and coincidentally settled in Dover, New Hampshire, after arriving on the fourth passage of the *Mayflower*. Thomas was a wealthy former urbanite who shared a bloodline with Geoffrey Chaucer, John Locke, and British academic artist Frederick Lord Leighton.[11] This information was passed from generation to generation via family folklore – but also was confirmed by genealogical research.[12] These family stories, though, have shaped the Leightons' collective identity, providing relatives who view this photograph with a means of connecting to a distant, ancestral past, and with the *Leighton Family* subjects. This generation of Leightons also would have been familiar with other ancestors whose personal histories intersected with the history of America. Jacob Leighton, Samuel's father, served as a surgeon for the Union during the American Civil War. Various Leighton relatives helped pilot the *Mayflower* voyages, served as soldiers in the American Revolution, and worked as constables in early settlements in Maine and New Hampshire.[13]

Nevertheless, this branch of the Leighton family presents itself as heirs to such a legacy of making history. They are a well-dressed farming and ranching family that enjoyed wealth and success as they settled the Oklahoma territory and engaged history's front lines. Specifically, the Leightons were among an assembly of settlers who paid the government for land seized from Native Americans at Kingfisher Land Office in 1994, after several land runs and lotteries brought more than 100,000 fortune-seekers to the area to attempt settlement. Only a few met successes. Many, especially non-farmers, left the territory before they even established a homestead for the required minimum of five years.[14] As late as 1899, historian Edward Everett Dale quipped that "one could ride west the whole day without seeing a homestead."[15]

Yet, in the poem "Ok-ar-che" (Fig. 3.3a [page 1] and b [page 2]), Miller notes that she was born in Dover in 1893, but nonetheless writes nostalgically of the land rush that occurred about four years before she was born. Her account, likely informed by family stories and folklore, celebrates the egalitarian nature of settling the frontier (somewhat blind to her family's white racial privilege as she does so). Miller writes:

> Race and color little mattered in this contest, free to all
> Rich and poor rubbed elbows, grasped a friendly hand;
> And they settled side by side,
> Shared grub boxes, did abide
> As true brothers in this fair and promised land.

Miller describes the frontier as a "land of milk and honey" with a climate that was "mild and sunny."

-Ok-ar-che-

1.

There was tumult and excitement in the spring of '89,
For those poised to make the dash for prairie home
News had spread like wildfire leaping,
Romance in young blood was seeping,
For Cheyenne-Arapahoe opening on Cimarron.

2.

Race and color little mattered in this contest, free to all
Rich and poor rubbed elbows, grasped a friendly hand;
And they settled side by side,
Shared grub boxes, did abide
As true brothers in this fair and promised land.

3.

Soon the blessings of all nature were showered on this territory;
Fragrance from peach blossoms, Heaven's own perfume;
It proved the land of milk and honey,
For the climate, mild and sunny,
Quickened fruit and field and prairie, all abloom--

4.

Cotton roses smiled demurely, near the fields of corn and wheat
Hope and cheer reigned now to free andilty,
Answering the willing hand.
Grasping boldly in this land,
Opportunities given only to the free.

5.

Thus they came and made their homes in Cheyenne-Arapahoe
Territory taken o'er by Uncle Sam--
In the path of Chisholm trail
Followed by Rock Island rails--
They fought elements and hardships, hand to hand.

6.

It was there that I was born near the little Dover town,
On the Cimarron in Eighteen ninety-three,
And I long to see once more
Familiar scenes and then live o'er
Childhood's fleeting mem'ries, now so dear to me.

Figure 3.3a, 3.3b Nancy Ellen Leighton Miller, "Ok-ar-chee" (Unpublished Poem, with Notes added by Samuel Bernard Leighton), n.d. Courtesy of Karen Leighton Belden.

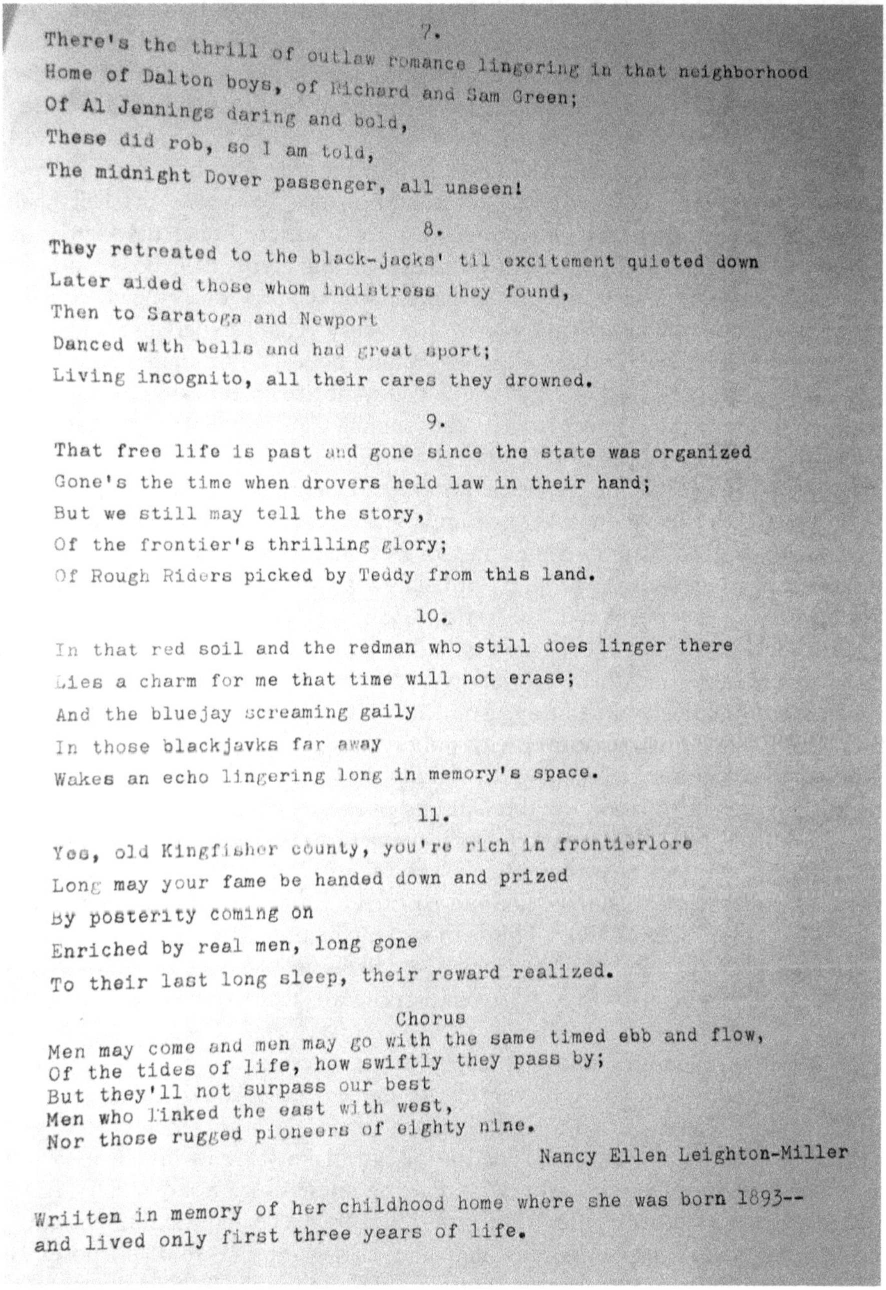

7.

There's the thrill of outlaw romance lingering in that neighborhood
Home of Dalton boys, of Richard and Sam Green;
Of Al Jennings daring and bold,
These did rob, so I am told,
The midnight Dover passenger, all unseen!

8.

They retreated to the black-jacks' til excitement quieted down
Later aided those whom indistress they found,
Then to Saratoga and Newport
Danced with bells and had great sport;
Living incognito, all their cares they drowned.

9.

That free life is past and gone since the state was organized
Gone's the time when drovers held law in their hand;
But we still may tell the story,
Of the frontier's thrilling glory;
Of Rough Riders picked by Teddy from this land.

10.

In that red soil and the redman who still does linger there
Lies a charm for me that time will not erase;
And the bluejay screaming gaily
In those blackjavks far away
Wakes an echo lingering long in memory's space.

11.

Yes, old Kingfisher county, you're rich in frontierlore
Long may your fame be handed down and prized
By posterity coming on
Enriched by real men, long gone
To their last long sleep, their reward realized.

Chorus

Men may come and men may go with the same timed ebb and flow,
Of the tides of life, how swiftly they pass by;
But they'll not surpass our best
Men who linked the east with west,
Nor those rugged pioneers of eighty nine.

Nancy Ellen Leighton-Miller

Wriiten in memory of her childhood home where she was born 1893--
and lived only first three years of life.

Figure 3.3a, 3.3b (Continued)

In reality, the Oklahoma frontier was rife with hardship and danger, and the idea that "[r]ace and color little mattered" surely would have been hotly disputed by the Native American tribes from whom the land was seized and given to white families like the Leightons. Newspapers reported regular skirmishes between Native Americans and settlers.[16] From 1885 to 1896, the area was hit hard by drought, and the land was considered so barren that "about the only sure crop was the rattlesnake."[17] Many settlers lived in sod houses or dwellings dug into the side of a hill, and had to fight against vermin, centipedes, and the constant presence of snakes in their shelters. Michael H. Reggio from the Oklahoma Historical Society noted that settler Andy Logsdon and his family "got so used to snakes crawling across their beds at night that they would 'just give a kick and they [snakes] would hurry on somewhere else.'"[18] Securing food and drinkable water also was not easy. Many settlers had to haul their water for miles in barrels, and ate bleached buffalo bones.[19] Fortunately, the Leighton family farm, acquired in 1894, was walking distance from the town of Dover, and included 160 acres of land punctuated by the winding Cimarron River, a valuable water source. This presumably gave the Leightons a better-than-average chance of success as homesteaders of the Oklahoma territory. Still, the Leightons did not stay permanently in Oklahoma. They left the land they purchased in 1894 in Oklahoma and in 1910 moved to a farmstead they purchased in Gove County in Western Kansas. Part of the Kansas farmstead remains the property of Leightons and their descendants today.

Miller notes in her poem that the Dalton Gang (which consisted of siblings Bob, Bill, Grat, and Emmett), whose parents had a farm about 10 miles away from the Leightons, brought "the thrill of outlaw romance" to the area, but also robbed passengers at the local train station. The Dalton Gang met its ends in 1892, a year before Miller's birth, in a shootout with police following a double-robbery gone bad in Coffeyville, Kansas. According to family folklore, the oldest Leighton son, Grover, provided farm labor for the Daltons after their three sons (Bob, Grat, and Bill) were killed, and while another Dalton son (Emmett) was jailed following the bank robbery.[20] Miller writes that much of the "charm" of outlaws and "frontierlore" came to an end when the state of Oklahoma organized governance and law enforcement and became a state in 1907. This is to say, by the time Miller grew to adulthood, the lawlessness of the pioneer days (which she likely would have been too young to remember with great clarity anyway, given that she was just six years old) had come to an end.

Such selective memories and desires to insert oneself into a story larger than oneself is an attribute of being human. Philosopher Paul Ricoeur suggests that such efforts to insert oneself in a broader temporal continuum are essential, "[b]ecause my personal identity is a narrative identity, I can make sense of myself only in and through my involvement with others."[21] Placing oneself into a larger narrative also is an act that expresses anxieties about mortality by exceeding the limited temporal scope of the human lifespan, while connecting to a greater sense of purpose. These Leightons, in other words, become part of the seemingly infinite story of their *Leighton-kind.*

Moreover, this urge to place *the personal* within a larger temporal continuum also can be seen as a reaction to industrial and technological modernity's continuing valuation of a global-minded, networked society, which extinguishes localized difference and individualism, Ricoeur also suggests.[22] The Leighton family witnessed the birth of one of the very first of such networks, the passenger rail system – or, the Rock Island Railroad. The institution of standardized "railroad time" and the time-and-space-"annihilating" railroad systems themselves sparked anxieties among the occupants of the corresponding towns that the community might lose its regional cultural identities, and their local specificity.[23] The Chisholm Trail – a symbol of the not-so-distant frontier past – was replaced in the 1890s by the "futuristic" Rock Island Railroad line, which provided the Leightons with expedient transportation to and from previous and future residences in Kansas and Iowa.[24] The Leighton family's neighbors, the infamous group of modernity-enabled opportunists, the Dalton Gang, robbed banks and trains, and conveniently boarded trains at the Kingfisher Depot near both families' homesteads, Nancy Ellen Leighton Miller recalls.[25] The Leightons' familiarity with the Daltons also gives them a connection to a broader historical narrative while simultaneously acknowledging their regional specificity and local individuality. The Daltons were infamous outlaws, but also were "neighborly," Miller notes.[26] Like the Leighton family itself, towns along railroad lines likewise struggled to maintain their specificity and identity while also participating in a regional technological network, the national railroad system.[27]

The Leighton Family as "Gumps": The Portrait-Taking Story

The poem "The Gumps's Pitcher Tooken" (Figs. 3.2a, 3.2b) tells the story of a busy, stressful day. The "youngans" – Ella and Hazel, who were age six and three, respectively – were "loaded up" on the wagon to go to town, and the older Leighton children "jogged along at quite pace, an soon they reached the town." Once the various Leightons arrived by wagon or on foot, they met up with "city-slicker" sister Elbye Ann – nicknamed "Phoebe Ann" in the poem – who already had married and left the Leighton home. Like her siblings, Elbye was proudly dressed in a brand-new gown for the occasion, and she appears spouse-free in this photograph featuring only the immediate Leighton family. Miller mentions in her poem that the boys were wearing brand-new squeaky shoes. This suggests that the occasion of having this photograph taken was a special event, and one that allowed the Leightons to indulge themselves by buying new clothing just for the occasion. Although this image bears no studio stamps or imprints, and its case or cardboard sleeve is not attached, it likely was made by a photographic studio located in the upstairs floor of a building in Dover, Oklahoma, a trading-post stop near the Chisholm Trail, a cattle/trading path leading from southern Texas to Abilene, Kansas.[28] Efforts to confirm Miller's account and determine which studio took this image were inconclusive. But, in the 1890s, studio photographer E. A. Carroll advertised his "Views of All Kinds, Made at Hard Time Prices" in the *Kingfisher Reformer*, and kept a studio practice in that area at the time the *Leighton Family* portrait was made.[29]

In her poem, Miller also takes the opportunity to point out the perceived social difference between rural-dwellers and urbanites – in her own family: "Now Phoebe Ann's nose tilted up, from livin' in the city, / An' the airs she put on then an' there, well it really wuz a pity." Although these comments may have been made in good-natured family humor, they indicate a persistence of a familiar stereotype of rural Americans as less sophisticated but more "authentic" and down-to-earth, and town-dwellers as arrogant. The clashing concepts of rural and urban living – and the stereotypes they evoke – is echoed by the language Miller uses in her poem, which differs dramatically from the articulate nature of her diction in the other poems.[30] Nevertheless, the sophistication of the poem's mechanics stand in contrast to its intentionally raw "country" language. Given that the Leightons were farmers living in the rural South and Midwest – yet were wealthy enough to purchase land rather than participate in a random raffle for it – the author's vacillation between sophistication and naïvete (even if it is mock naïvete) is appropriate without appearing insincere. Farming is a dirty, labor-intensive way to make a living, even for those who make a good livelihood in the vocation. The Leightons were a middle-class family with access to solid schooling, and that could afford "luxury" items such as dress-up clothing, family portraiture, and newspaper subscriptions.

Miller self-deprecatingly calls her family "the Gumps," after the everyday, relatable, urban, middle-class characters in a popular newspaper cartoon strip drawn by Sidney Smith from 1917 to 1959, and nationally syndicated in 1919.[31] The serial, like most comic strips, wove narratives that unfolded from frame to frame, with interdependent drawings and text. It provided a model for storytelling for the mass public. Smith's plots communicated a clear statement of American values that lauded hard work, individual initiative, and looked to the family as a source of social order.[32] This reference indicates that the earliest year Miller could have written this poem was 1917, eighteen years after the photograph was taken. Miller's memories were thus at least two decades removed from that day's events by the time she committed them to paper. They also were influenced by her reading of "The Gumps" comic strip, which told the stories of an ordinary city-dwelling family consisting of a father (Andy), his nit-picking wife (Minerva/"Min"), their two sons (Sam and James), their two pets, and their housemaid (Mary). Andy was loud, loving, optimistic, and somewhat blind to his own faults.[33] He is prone to seeking out one get-rich-quickly scheme after another. Andy is kept in check by Min, who grounds the whole family in the values of hard work while also upholding her traditional gender roles as homemaker and mother.[34] While Min is the disciplinarian of the two, she never attempts to usurp Andy's power as the head of the household, which would rock the established social order. The Gump children and Andy view Min as the somewhat deflating voice of reason, and they mock her openly.[35] "The Gumps" were a mass-culture phenomenon, and inspired spin-off projects including film shorts, a radio show, and the merchandising of toys and household items. They became a touchstone for the middle-class American family, and an inspiration for readers to find humorous narratives in everyday family life.

Framed by the title "The Gumps's Pitcher Tooken," Miller's poem describes a family that is endearingly imperfect, with personality dynamics that are not unlike that of the fictional comic-strip "Gumps." She writes that on the morning the photograph was taken, bitter cold weather caused the children's noses to "run like sap," turning the Leightons' noses red. Mother Cynthia's reportedly commented that "'twouldn't matter, fer red noses wouldn't show, / In the pitcher, whether red er yeller, or as white as snow." This illustrates that Cynthia had an understanding for how black-and-white photography would "flatten" red and yellow hues into similar shades of the gray-scale. That is to say, Miller (and perhaps even Cynthia) had been conditioned by the prevalence of photography in rural American culture by 1899 to understand and anticipate its peculiar form of translating the color of "lived" experience into photographic black and white.

The exposure time's precise temporal span remains elusive, although the introduction of gelatin emulsion in 1878 allowed for exposures as brief as 1/25th of a second, and this duration was decreased to 1/250th of a second by 1900 – one year after this image was made. In spite of her attention to recreating a narrative of the day the *Leighton Family* portrait was made, oddly, no specific date of origin is written on the image or provided by Miller. In a typical year in Oklahoma, the weather in the months of January, February, November, and December can get cold enough to redden cheeks and make noses run. Miller's poem thus gives the undated photograph a seasonal time frame. Because the poem mentions Lucinda's upcoming "fall" wedding, this photograph likely was taken between January and February of 1899. The poem's reference to weather – which is, by nature, cyclical and contingent – allows for greater specificity in placing this photograph in a temporal continuum, while positioning the image in natural, meteorological seasonal time.

But just to be sure that the photograph did not capture any tell-tale marks of their frigid journey to town for posterity, father Samuel Humphrey (whose sense of humor is represented as quirky, like that of Andy Gump) insisted that the children put cornstarch on their faces. This is a source of levity in the poem: "When they finished with their primpin', Pa laffed an' sed, By gorry / If they didn't look like plaster paris busts of Queen Victoria." Miller's prose is peppered with good-natured jokes and nostalgic, humorous stories about events, such as the "cornstarching" of the Leightons' red noses, while also making a sophisticated reference to portrait busts of the Queen of England. Her poem's account of that day is consistent with the conventions of family photography. In them, we only see the happy moments. Very seldom do snapshots commemorate the everyday routine, or incidents of hostility.[36] In her poem, Miller describes the day as quirky, fun, and chaotic. She remembers it and her father's good humor fondly. Miller sees the Leightons as "Gumps," and the act of preparing eight children for a photographic portrait as an adventure rife with big-family chaos.

When the Leightons reach the photography studio, Miller describes their journey of climbing up some stairs. Samuel David ("Saul David" in Miller's poem) fell head-first down the stairs. Miller takes a good-hearted jab at her older brother Samuel's clumsiness in her poem: "It almost knocked his senses out, tho' I reckon he hadn't much in him, / Fer all he ever wuz good fer wuz a blabbin' an' a grinnin'." The poem then describes stern mother Cynthia's reaction to Samuel's fall. She "boxed his ears an' sed, / 'Git yerself together now, er you'll go home to bed.'" All of the Leightons pitched in to help dust off Samuel, fighting the urge to laugh about his fall, and assumedly, to avoid getting reprimanded by Cynthia. The shared experience of growing up with a strict mother united the Leighton children, Miller suggests. Cynthia, according to the poem and other oral accounts, was strict, domineering, and in charge.[37] She also controlled Samuel Humphrey – who is described as good-natured and jovial by Miller – much as the stern disciplinarian Minerva of "The Gumps" exerted power over family patriarch Andy, as well as everyone else.

When the Leightons entered the room and lined up to be photographed, Miller notes that they looked "purty as you please," but that they were uncomfortable with being pushed so closely together to pose. As the camera's shutter was tripped, Miller notes that all of the Leightons – except, presumably, for mother Cynthia – wanted to laugh out loud after watching Samuel David fall. However, because smiles would not be fashionable in family photographs until Kodak marketed the convention in the 1930s, "they couldn't laff atall."[38] Meanwhile, Samuel David did not share the group's sense of humor over his fall. His face "wuz puckered tight, frum gettin' boxed so good." Indeed, Samuel David wears a peculiar, puckered expression in the portrait as he stands right next to his mother.

At the moment the shutter was open, Miller's poem notes that a fly landed on Samuel Humphrey's nose "when he couldn't bat an eye," because of the time required for the camera's exposure. Indeed, a small dark speck is visible on his right nostril. It is centered on a raised circular-shaped, flat area that more likely could have been a bandage with a spot of breakthrough bleeding, a buildup of scar tissue from an injury, or even an unusual facial cyst, wart, or another growth. Samuel Humphrey is fortunately situated to the left side of the image, where his right nostril would be in shadow as he stares at the camera. Nonetheless, the fly story – along with other details provided by Miller – enliven the photograph, giving it a sense of temporal vibrancy. Samuel David's tumble occurred in the immediate, "off-camera," *past*, and was fresh in the minds of the photograph's subjects, during the moment we see indexed by the image. The "fly" that landed briefly on Miller's father's nose (or, perhaps, his misremembered injury/facial deformity) enjoys a *suspended present* tense, frozen eternally in its position in the *past* (even though the possibility of a fly surviving a cold Oklahoma winter is quite unlikely).

Miller's brother's unfortunate plummet also made it difficult for most of the Leightons (except for Cynthia and Samuel David himself) to remain properly straight-faced for the photograph to be made. Hints of a smile are apparent on Isa Belle's face, but on no one else's. The subjects' struggles to control their reaction to Samuel's fall are frozen by Miller's poem and in the photograph – in an eternal, enlivened dureé of *present* tense that perplexingly lies in the *past*. The image provides a visual account of ten people suspended (supposedly) in an infinite temporal duration of emotional restraint – of laughter, anger, and annoyance. However, Miller's poem, which was written at a time when smiling was a convention in family photography, may be an effort to recast the stern-faced Leightons as lively and jovial. Instead of remembering an emotionally distilled childhood implied by the straight faces in this image, Miller provides an animated narrative that affirms family bonds, and expresses humor and great fondness. Miller provides an alternative narrative for a family that would have been devastated from the loss of two children – one as an infant, and another at age four – while they were homesteading in Oklahoma.[39]

The Subjectivity of "Narrative Time"

On the day the *Leighton Family* photograph was taken, Miller was about six years old. She was not then capable of writing the poem, and would not do so for several decades. Her account of that day is dependent upon her memories – which were, as *all* memories are, selective, biased, perpetually evolving, and have a tendency to deform reality.[40] Her memories were supplemented by family folklore and recorded when she was an adult. Psychology scholars Michael Ross and Anne Wilson have argued that personal memories and histories evolve to reflect parallel developments in belief, experience, knowledge, and changing goals.[41] Miller's view of her childhood is tinged with retrospective nostalgia.[42] French historian Pierre Nora argues that histories such as Miller's only are capable of operating in opposition to "objective" histories because memory is malleable.[43] Miller's history, while striving to evoke historical authority, indeed is inescapably ridden with subjectivity, omission, and romanticization.[44]

Anxiety over the disappearance of a preferred, participant-provided, first-hand history – *her* history – for the continuum of time surrounding the *Leighton Family* snapshot drove Miller to place the image in her subjective, creative, "narrative time." Her story grants the image temporal specificity, rescuing it from the symbolic and placing it in the realm of the particular. As Geoffrey Batchen argues in *Forget Me Not: Photography and Remembrance*, family photographs function to express the individual uniqueness and importance of a human – efforts he argues are motivated by a Barthesian death anxiety, and the desire not to be forgotten.[45] Such images also function for their narrators as a means of affirming *their own* concepts of self. For example in *Photo Album of Family D., 1939–1964*, contemporary French artist Christian Boltanski borrowed family photographs from a friend, re-photographed them, and attempted to put them in chronological order. He devised an accompanying fictional chronology for the subjects of these photographs, making an implicit reference to the fictional narratives snapshots provoke. This urge to impose linear, chronological order upon the images illustrates the desire to create an orderly narrative, and the degree to which such an approach to storytelling has become a convention. Boltanski explains: "Knowing nothing of these people, I set out to reconstitute their lives on the basis of photos taken at all the important moments of their existence, which should remain as testimonial to their lives after their deaths." He explains the resulting conclusion of the project: "I realized that these images were only witnesses to a collective ritual. They didn't teach us anything about the Family D … but only sent us back to our own past."[46] This is to say, perhaps Miller's narrative for the *Leighton Family* says more about Miller than it does about the photograph.

In her poem "The Gumps's Pitcher Tooken," Miller further describes the occasion for the group photograph, the *then-future* nuptials of "Lucindy." Records confirm that Lilly Lucinda Leighton married Henry Schaefer on December 24, 1899, at the Schaefer farmhouse in Wayne County, Iowa.[47] The wedding was not scheduled for "this fall," as Miller writes. The poem, a work of prose based on Miller's memories of events that transpired when she was six years old, combined with second-hand memories shared with her by other family members, thus abandons the pretense of objectivity. Nora argues that our memories of events are inherently flawed, lending subjectivity to even histories that claim to be objective:

> Memory … remains in permanent evolution, open to the dialectic of remembering and forgetting, unconscious of its successive deformations, vulnerable to manipulation and appropriation, susceptible to being long dormant and periodically revived … Memory, insofar as it is affective and magical, only accommodates those facts that suit it; it nourishes recollections that may be out of focus or telescopic, global or detached, particular or symbolic – responsive to each avenue of conveyance or phenomenal screen, to every censorship or projection.[48]

However, instead of aspiring to create an "objective" account, Miller's narrative functions to help form, affirm, and maintain self-perceptions, conceptions of fond familial relationships, romanticized conceptions of a struggle-free childhood, and notions of the roles and personalities of different members of the family. While these constructions are indeed rooted in folklore and conventions of self-presentation – and therefore are anything but objective and "truthful" histories – this matters very little. To detach the personal and the subjective from narratives for this image would rob the portrait and narrative of what makes them into a living piece of personal folklore, which is shaped, molded, fashioned, refashioned, enriched, and invented by the malleable human memory – and which, according to literature scholar John Paul Eakin, "is one more source of fiction."[49]

Siegfried Kracauer further argues that expectations of the "facticity," or "truth value," of a photograph differ from expectations of a temporal "memory image" – in which the viewer selects only the things he/she wishes to retain. According to Kracauer, a photographic image keeps everything it sees, but the memory image selects what it sees, in the *mind's* eye.[50] Personal memory therefore makes the photograph's status as an objective document of secondary importance – as long as the image's photographic "facticity" does not overwhelm the personal mythologies and narratives associated with the temporal image. Miller's poem consciously or unconsciously participates in the modern project of disassembling master "objective" narratives, and experimenting with new, subjective approaches to storytelling.[51]

Time Anxiety and the *Leighton Family* Portrait

As Miller suggests in her poem, anxiety about the dispersion of the Leighton family unit prompted the taking of this photograph: "So, before the family circle wuz busted up fer good, / They loaded up the youngans to go get their pitchers took." This photograph's function *was* (and still *is*) to affirm individual and group identity by providing a sense of familial cohesion. The children pictured in this photograph would all end up marrying, and they lived in areas such as Colorado Springs, Colorado, Modesto, California, Dover, Oklahoma, Wayne County, Iowa, and in cities all over the state of Kansas. While Cynthia and Samuel Humphrey Leighton are not finished rearing children, and would give birth to one more surviving child in 1900, the photograph is marked by the implications of biological time through references to the "life cycle" and Cynthia's child-bearing "biological clock."[52]

Shortly after the group portrait *Leighton Family* was made, daughter Lilly left the family home to marry Henry Schaefer and live in Wayne County, Iowa. Industrial modernity's relatively rapid new forms of transportation, such as the Rock Island Railroad, reordered familial relationships by enabling offspring such as Lilly to disperse geographically to start her own family.[53] Anxiety over the future absence of Leighton offspring fueled the employment of a photographer to capture past moments of family togetherness in the *Leighton Family* portrait. Its further implication of an oncoming marriage, an event of the encultured "life-cycle," places the Leighton family in the throes of change, and within the fleeting, relentless passage of time.

The photograph is itself an attempt to freeze time, to halt its unyielding passage. This collision of past and present and future provided what Barthes called photography's "madness," its ability to offer "an image of the real, of something that *was*, that we can hold in our hands … but that does not physically exist in our time and space."[54] The *Leighton Family* portrait similarly evokes a Barthesian *future-anterior* tense. The photograph exists to quell unease about changes to the family unit that will occur in the future. Consequently, the image speaks to the fact of time's passage, and to the subjects' eventual and inevitable deaths.[55]

But the photograph is not only a product of temporal anxiety over the advance of the "life cycle." The cumulative sensorial effect of the invention and popularity of rail travel, telegraphy, and telephony – all of which had been invented and assimilated into American culture by the time the *Leighton Family* portrait was made – also produced anxiety. The Leighton parents decided to make one last group photograph, knowing that their three oldest children were nearing the age at which marrying was the norm. Their looming departures were a threat to perceived family cohesiveness. The subjects of the *Leighton Family* lived in an age that redefined time and space. As cinema scholar Mary Ann Doane suggests:

> At the turn of the century time became palpable in a quite different way – one specific to modernity and intimately allied with its new technologies of representation (photography, film, phonography). Time was indeed felt – as a weight, as a source of anxiety, and as an acutely pressing problem of representation. Modernity was perceived as a temporal demand.[56]

Miller has good reason to be concerned that her story could be forgotten. As time's passage progressively widens the gap between the image and its subjects, and number of people who might know of its associated family folklore diminishes with every passing day, the subjects of the *Leighton Family* come closer to operating as symbols themselves, as they gradually lose the personal specificity granted to them by individualizing family narratives. Instead, the Leightons become "types" or "characters" who evoke subjective associations in viewers. As decades pass and references to the "Gumps" even fall on an audience unfamiliar with newspaper-syndicated comic strips, even those "characters" are detached from their context. When albums or family portraits such as this one appear free from contextualizing material in museums and galleries, viewers provide new narrative contexts for the Leightons' lives and plights. Time's passage and the diminishing knowledge of the image's temporal specificity increase the *Leighton Family*'s symbolic potential, and the possibility of the Leighton family's individuality being lost forever.

Yet, insomuch as *every* recollection of the past is an active, reconstructive process – one which involves subjectivity – theoretically, every historical narrative is revealed to be perhaps just as subjective as, for instance, Miller's narrative.[57] This debate, which is currently of keen interest in memory discourses, dissolves the boundaries between "objective" and "subjective" histories, destabilizes notions of an "absolute truth" and undermines the project of "master" narrativity.[58] Some narratives, historians might agree, are preferred to others. But do such concerns about narrative quality and truthfulness matter, given the family photograph's function as an object which aids in the building and maintenance of self-concepts, personal memories, and histories by prompting the creation and maintenance of subjective narratives?

Observers' Reinterpretations of the "Exhibited Self"

The *Leighton Family* portrait provides further convolutions of temporality that exceed Barthes's future-anterior tense. In a caption for the photograph *Leighton*

Family, written by "narrator" Samuel B. Leighton (son of portrait subject Grover), the absence of a family member is noted. William Orman ("W. O.") Leighton was not yet born at the time the photograph was made (Fig. 3.1b). By giving *Leighton Family* (1899) a caption that refers to a *future* event, William's birth (on October 28, 1900), Samuel reframes the photograph by imposing a presaging *future* presence and narrative voice (from around the year 1980) to speak for this image. The *Leighton Family*, as narrated by Samuel B. Leighton's caption, invites viewers to transport themselves, temporally, into the S. H. Leighton family – which does not know that it will welcome a baby boy a year later.[59] The caption provides an upbeat reversal of the Barthesian expectation that every snapshot foreshadows *death*. However, viewers who center *themselves* – rather than members of the S. H. Leighton family – as the subject of consideration may find death anxieties peaked by the eternal youthfulness of its subjects, appearances which mask the fact that all of the participants in this photograph have been deceased since 1988. The same fate awaits all of us. Another grim reminder of the looming specter of death – with profound influence upon *those who exist* in the present – in the *Leighton Family* photograph is the absence of four children, three of whom died at birth, and another died at age four of unknown causes.[60]

Another example of the highly personalized recasting of memory and personal history is apparent in a revised version of the *Leighton Family* photograph. An untitled montage made by William Orman Leighton's daughter, Billie Hazel Leighton Flora (b. 1921, d. 2012), in the 1980s (Fig. 3.4) completes the *Leighton Family* circle by including its *then-not-yet-born* (but *now*-deceased) member. In Flora's version of the *Leighton Family*, a photograph of William Orman (b. 1900, d. 1987) appears as a toddler wearing a white dress. He has been scaled to a proportional size by photocopying, cut out with a pair of scissors, and was pasted into a prominent, free-floating, front-row space of the *Leighton Family* portrait, along with the other young Leighton siblings. He "sits" on the imaginary left knee of his mother and appears to be about the same age as sibling child Hazel, then age three or four. The physical resemblance between Hazel and William is uncanny.

Flora's montage implies her regard for the photograph as a "document" that would be viewed by posterity as a testimonial to the full cast of offspring in *her father's* generation of Leightons. Given the option of inserting her father's presence in a caption, as Samuel B. Leighton did for William, Flora instead chose to insert a *photographic* record of his existence. William is thus granted equal visual and descriptive presence to his siblings. Viewers may come to know William's appearance, its relationship to that of his siblings and parents, and, possibly, its similarity or difference to their own.[61]

Flora does not disguise her photographic montage's synthetic nature. The edges of the image of William are jagged and reveal the white edges of the photographic paper from which it was cut. A caption reading "Added W. O. Leighton born 1900" announces Flora's fictional visual addendum. Seamless naturalism is not Flora's goal, though. Instead, what matters is that she "correct" the potentially misleading "truth" to which the photograph was witness by illustrating – photographically, and for posterity's sake – that the Leightons had a total of *nine – not eight* – surviving children, and that her branch of the family tree was a part of the family narrative, too.

Literary theorist Paul A. Harris, who studies the subjectivity of memory as a narrative, suggests that the "constructedness" of our memories becomes "starkly

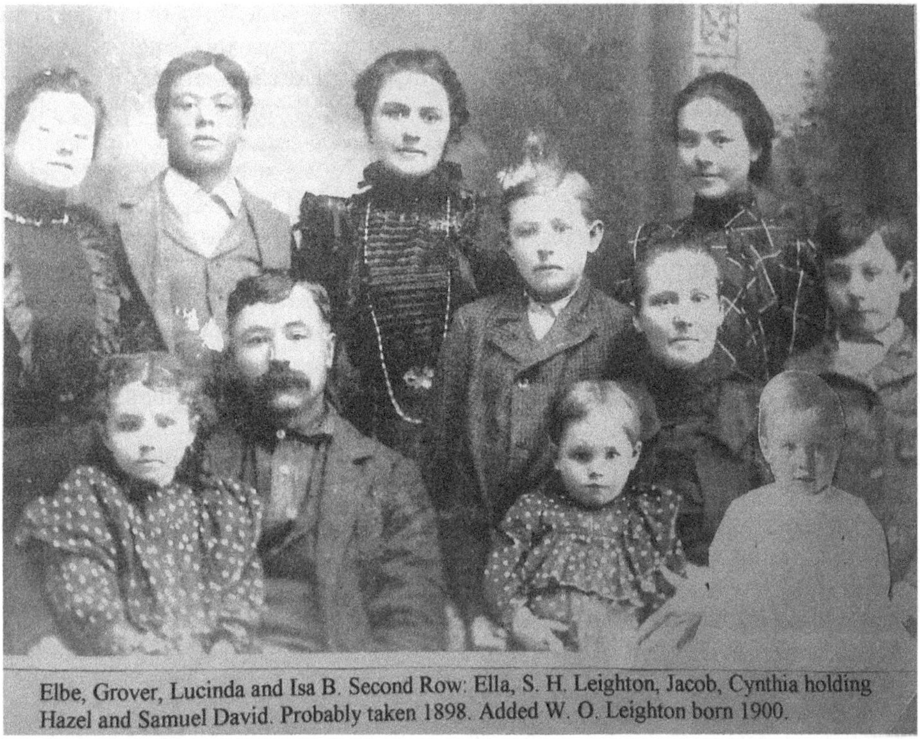

Figure 3.4 Photographers Unknown/Billie Hazel Leighton Flora, *Leighton Family*, c. 1980, photocopied photomontage, 4½ in. x 6½ in. (11.4 cm x 16.5 cm). Courtesy of Karen Leighton Belden.

clear" when our narratives are compared to an "external memory system."[62] This is to say, an image such as the *Leighton Family* conflicted with Flora's narrative of her father's – and by extension, her – existence. To "correct" the photographic "external memory system" (that Flora anticipated would outlive her internalized biological memory), she introduces the "then-future" of her father's existence to the image, more than eight decades after the *Leighton Family* was made. In her montage, time stretches backward, then forward, and back even farther, adding yet more interpenetrating temporal textures to the group portrait.

Albums, snapshots, and the stories that accompany them, Martha Langford suggests, are a means by which family history and identity are passed down from generation to generation.[63] Flora, who died in 2012, was a history buff and the organizer of the annual Leighton family reunion. She was deeply invested in the keeping of family history and folklore. If her father was not present, photographically, he might be left out of the recitation of family history – and so might she, and any of her present and future bloodline. Flora's revised image cheats the limitations of the linear "lived" time of the photograph while also expressing her anxiety over being forgotten, in the future, after she has died and can no longer assert her own presence in the family in the future.

Her montage conveys a need to connect to a broader span of time than the one naturally afforded to her by a lifespan. By pasting her father into the image, Flora anticipates her image will reach future audiences, implicitly expressing her anxieties about life's ephemerality and finite temporality. But Flora's creation of the photographic montage, which is stapled to a detailed hand-written family tree, also is motivated by the anxiety that later narrators of the *Leighton Family* portrait might assert narrative control and, as Hayden White argues, recast history as *they* see fit.[64]

By adding her father – as a toddler – to the image, Flora reconnects a portrait of William to the context of his childhood in an attempt to reverse the linear generational temporal order and recontextualize his life. Yet, in spite of photography's capacity for temporal flexibility, Flora can never fully recover any new knowledge of her father as a child from looking at the image – even within its new expanded context as part of the *Leighton Family*. In this sense, photography frustrates its makers, keepers, and narrators. In a way, Flora's effort is like that made by Barthes in *Camera Lucida*, as he takes a closer look at photographs of his then recently deceased mother in an attempt to get to know more about her as a child – rather than as the adult, maternal figure that lives in his memory. His attempts to gain knowledge about her from only photographs are ridden with complication, and he ultimately deems them unsuccessful.[65] While the temporal complexity of photographs tantalizes us with the impossible visual experience of an intangible, inaccessible space of time, the image denies its observers true time travel. Photographs cannot allow Flora and Barthes to get to know their deceased parents or to truly transcend time's relentless march.

Nonetheless, Flora's photomontage activates several tenses, simultaneously. It makes reference to the discordant *past tenses* of the exposure times of both photographs, while also indexing Flora's act of revising the image in the 1980s. This montaged *Leighton Family* portrait adds Flora's *"then-future-but-now-past"* narrative voice, and the implication that she is motivated by honoring her father's memory and continued memory by future viewers of the photograph.

The *Leighton Family* Portrait in Time/s

This chapter's analyses of the *Leighton Family*, as well as Flora's photocopied montage of the *Leighton Family* portrait with her father added to it, have uncovered multiple temporalities and simultaneous *narrative times*, including: (1) *ongoing, genealogical time*: the temporal continuum of the pioneering Leighton bloodline; (2) *master-narrative time*: the Leightons' engagement in the "master-narrative" story of the founding and settlement of America; (3) the *instantaneous duration of past time* in 1899 in which the photograph's exposure was made; (4) the *"now-past-but-then-future"* tense of William Orman's birth in 1900, and Lilly Lucinda's marriage and departure from the Leighton family in late 1899; and (5) Samuel Bernard Leighton's and Billie Hazel Leighton Flora's narrative voices, which were imposed upon the image in the 1980s to give William a place in the image – and in *time*; (6) *Barthesian future-anterior time*: the inevitable death of the subjects of the photograph, and their possible *Ricoeurian, eternal time* through the photograph's re-narration; (7) *biological time*: each of the subject's position in the "life cycle"; (8) *observed biological time*: points of entry for a viewer's possible identification with different subjects' positions in the "life cycle"; (9) the absence, *both "then-past-and-now-past" and "then-future-but-now-past"* of four children who died in infancy and at age four; and (10) the *now-absence* of the lineages of these children who did not survive to adulthood. The *Leighton Family* temporally indexes or prompts consideration of each of those points in time, evoking non-linear, multiple temporal vacillations between past, present, and future tenses (and combinations of these). This exemplary portrait has a complicated, unfixed, and potentially endless number of ways to relate to temporality that far exceed Barthes's conception of future-anterior time.[66]

"Narrative Time" and Subjectivity: Precedents in Practice and Theory

Observers' tendencies to indulge themselves in subjective departures from the "facticity" of photographs when they narrate images also appears in the previous commentaries of both Walter Benjamin and Barthes. In "A Little History of Photography" (1931), Benjamin identifies the woman in the wedding portrait *Karl Dauthendey (Father of the Poet), With His Fiancée* as Dauthendey's first wife who committed suicide.[67] As Carolin Duttlinger points out, the woman in the image is his *second* wife.[68] Benjamin does not acknowledge or discuss the error in the text. Duttlinger suggests that this mistake presents "an alternative concept of aura, one which transcends fixed historical or technological categories through the model of an imaginary encounter between viewer and image."[69] Benjamin's error exemplifies viewers' tendencies to create multiple possible narrative contexts for photographs that convey varying truths.[70] These variegated imaginative anecdotes detach the photograph from any singular auratic experience in a fixed time and space. Instead, these narratives offer the possibility of an *infinite* number of temporal experiences.

Margaret Olin has argued that Barthes made similar, perhaps deliberate, "mistakes" in *Camera Lucida*. His inaccurate identifications of details from several photographs in *Camera Lucida* also make implicit claims about the dubious nature of an objective narrativity. Olin points out that Barthes describes a *punctum*-provoking necklace in a discussion of a photograph by James van der Zee that did not exist in that image, but instead appears in a photograph of his Aunt Alice, which is published in a different book, *Roland Barthes par Roland Barthes*[71]:

> This mistaken detail, then, not the necklace actually pictured, led Barthes to the center of pain in the photograph ... Indeed, the wearer of the necklace, Barthes's Aunt Alice, occupies the same place as Van Der Zee's "solacing Mammy" in the family picture, or at least in the picture of the family, and the composition of a photograph, not the pumps or the necklace on a real person, enabled him to make the identification. Presumably, Barthes recognized the family constellation, even though to do it he had to move the detail, the *punctum*, from one photograph to another.[72]

Although his transplanting of a *punctum* from one photograph into another could have been an oversight – and his untimely death leaves his intentions forever unknown – Olin argues that this gaff was purposeful.[73] Barthes's "mistake" implicitly suggests, intentionally or unintentionally, that viewers of photographs project their subjective, constructed, often fictional, narratives upon photographs, without particular concern for the images' content. The necklace – an object that never had a presence, or a *that-has-been*, in the described photograph – provided the image's *punctum* for Barthes, giving the image with its personal meaning, which varies by viewer.[74] In other words, temporal specificity is portable, mutable, and is open to imaginative interpretation and re-interpretation.

Like the gold-ribbon necklace Barthes "sees" in van der Zee's portrait (but which actually appears in one of Barthes's family photographs), Miller's poetic narrative for the *Leighton Family* conveys an anxious projection of *the personal* and specific (albeit an inaccurate, hindsight history) for an image that was becoming increasingly symbolic and detached from those who might breathe new life into it by narrating it as family folklore.[75] Flora's photocopied montage of a photograph of her own father onto the Leighton group portrait likewise projects an external *punctum* – her father's early presence – into the portrait. A family photograph's tendency to be a site for such projected subjectivities, gaffs, and omissions challenges claims that the photographic image is an objective document.

Moreover, Olin and Diana Knight provocatively argue that Barthes's discussion of a photograph of his recently deceased mother at age five – an image which never appears in *Camera Lucida*, and has yet to be discovered – may be based on a non-existent image.[76] Nonetheless, Barthes vividly describes the subject of the Winter Garden photograph as follows:

> My mother was five at the time (1898), her brother seven. He was leaning against the bridge railing, along which he had extended one arm; she, shorter than he, was standing a little back, facing the camera; you could tell that the photographer had said, "Step forward a little so we can see you"; she was holding one finger in the other hand, as children often do, in an awkward gesture.[77]

Olin points out the similarities between the hand gesture he describes and the manner by which Barthes himself holds his hands in a snapshot reproduced in *Roland Barthes par Roland Barthes*.[78] She also notes similarities between the Winter Garden photograph and an image of Franz Kafka which was described by Walter Benjamin in "A Little History of Photography" and appeared in *Le Nouvel Observateur*:

> But most likely there was no Winter Garden Photograph to reproduce, or perhaps only the one of Franz Kafka at the age of six, described, with its palm trees and Kafka's soulful eyes, as well as an oversized hat, by Walter Benjamin in his essay "A Short History of Photography." Benjamin placed the setting tentatively in a "kind of winter garden landscape," but the French translation that appeared in *Le Nouvel Observateur*'s special issue on photography places Kafka definitively in a winter garden.[79]

His analysis of this image – rather, his placement of it within a *narrative time* – is an effort to cheat linear time and get to know his mother as a child, according to Barthes.[80] However, Olin argues that his efforts inevitably lead Barthes back to a consideration of *himself*:

> In the Winter Garden Photograph, Roland Barthes discovered not his mother, or not only his mother, but also himself, himself as a child, specifically as a child known from photographs. A chain of photographs leads Barthes, searching from

> image to image, to the unexpected discovery of himself as his own mother, just as he had been his mother's mother while he cared for her during her last illness.[81]

Olin here suggests that viewers project *their own* narratives into photographs, regardless of the image's unique exposure-time-oriented narrative context. Viewers' autobiographical readings thus force a dialog between the private and the public, and between micro- and macro-histories. Jan Campbell and Janet Harbord suggest that Benjamin's work can be seen as a means to explore "connections between consciousness and the world, collective and individual history."[82] Not only is the insertion of autobiography a legacy of the work of Benjamin and modernism, but it also is purely postmodern in its embrace of plurality.[83]

The story of the Winter Garden, Olin suggests, is really an encounter with the temporal "otherness" of his own childhood, which he knew partially from photographs. That is to say, for Barthes, the analysis of the Winter Garden photograph is driven by the phenomenon of knowing oneself *through photographs*, rather than from the memory of experiencing the referent slice of time during which he was *photographed*. After all, Barthes's mother might not have remembered herself at that age, either, and similarly constructed her notion of "childhood" from images such as the Winter Garden photograph. What results, in part, from Barthes's discussion is a preponderance of the modern phenomenon of knowing oneself via photographs. Moreover, as the image becomes a springboard for the projection of the viewer's associations, the existence of the Winter Garden image is unnecessary. As a family portrait from a commonplace genre of photography – like the *Leighton Family* – we can imagine the Winter Garden image without needing to see it.

A future-anterior anxiety underlies Barthes's discussion of the Winter Garden photograph. He regards his mother's image and ponders not only her recent death, but also the inevitability of his own. However, Barthes's suggests that additional temporal dimensions exist in his analysis. The childhood portrait gives material visual form to a span of time experienced (by his mother), but perhaps not remembered by her – except through photographs. Barthes further describes the Winter Garden photograph as paradigmatic of *all* photographs, claiming that "[s]omething like an essence of the Photograph floated in this particular picture."[84] Barthes's Winter Garden photograph – the one image that speaks for them all – invites amalgamations, projected memories and shifting observer-generated narrative times.[85] The medium's power, Barthes implies, lies in the multiple, temporal associations it evokes in the observer.

Vernacular History and "Objectivity"

In 2003, Samuel B. Leighton – the son of Grover, the eldest son in the *Leighton Family* portrait – recorded an oral history for a studio-made portrait of himself as a young boy (Fig. 3.5a [front] and 3.5b [back]). His story presents a unique opportunity to further study the omissions, additions, and motivations behind the narratives provided by subjects to speak for posterity about photographs.

In the portrait, Leighton smiles as he sits in an oversized wooden chair with his arms around a long-haired white dog and an oversized teddy bear. This commonplace example of an early-twentieth-century photographic portrait was taken at

Raphael Studio in St. Louis, Missouri, in 1915 or 1916. The image is reproduced on a postcard, and printed within an inverted, tilted, whimsical apple silhouette. Leighton wears a crisp striped sailor's shirt, matching shorts, an undershirt, and calf-length socks. Rather than wear the typical rural-Midwestern school-age child's wardrobe of either overalls or a suit, Leighton's sailor garb, he claimed, distinguished him as a fully-assimilated St. Louis urbanite.[86] He meets the camera's gaze with squinted eyes.

The letters "S. B. L." – for Samuel Bernard Leighton – are written in the bottom right corner of the postcard in Leighton's adult handwriting. On the back of the print, Leighton also added these notes, which are hand-written in blue ink: "Me? / This was Taken / in St Louis / I still Remember / The White Dog / He was Cross / S B L." Additional notes appear below, written in a lighter blue ink, presumably during a different session of viewing the photograph: "Dad and I / lived on Market St /St Louis, Mo / 1914–1916 / Then came / Back to Kans / 1916 S.B.L."

In these hand-written notes, Leighton affirms as "fact" something to which the Raphael Studio imprint attests: the portrait was made in St. Louis between 1915 and 1916. (Although Leighton dates the photograph as 1914–1916, Raphael Studio would not open for business until 1915.[87] So the date of this photograph's taking

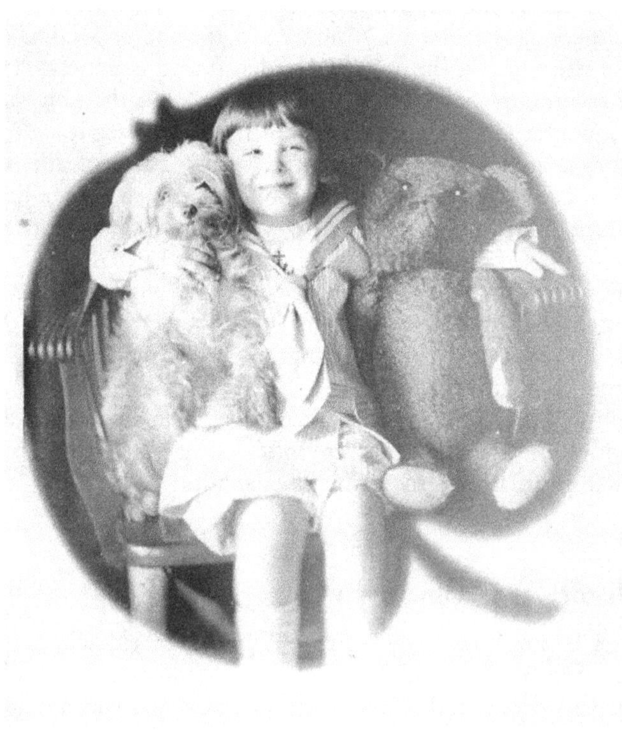

Figure 3.5a, 3.5b Raphael Studio, St. Louis, Missouri, *Untitled Portrait* (front and back), c. 1915–1916, albumen print on postcard stock, 3 in. x 5 in. (7.6 cm x 12.7 cm). In Figure 3.5b the handwritten notes are by Samuel Bernard Leighton. Courtesy of Karen Leighton Belden.

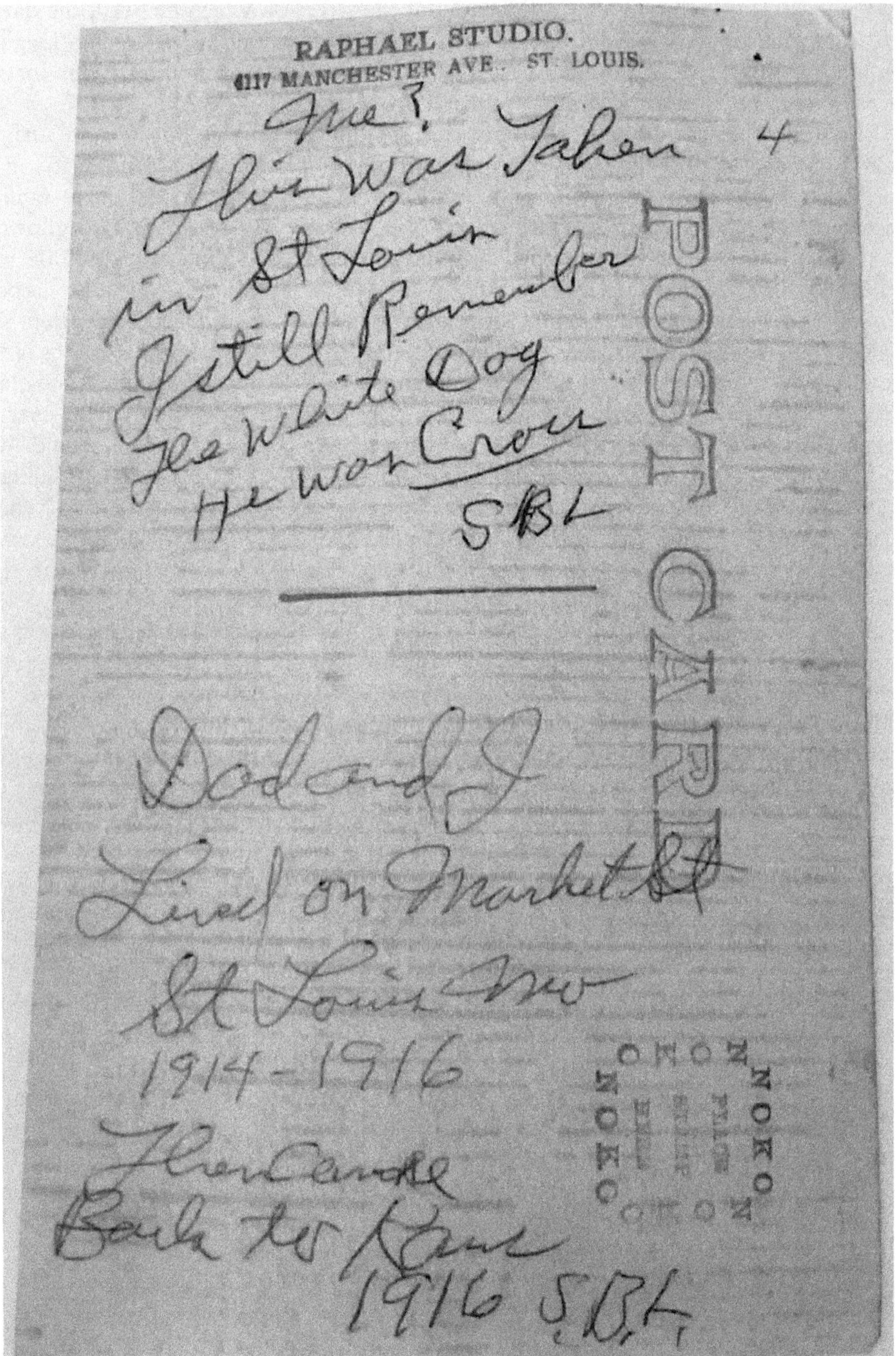

Figure 3.5a, 3.5b (Continued)

can be narrowed further to 1915–1916.) In the oral history he provided for this image in 2005, Leighton recalled the name of the street on which he lived, the dates of his residence there, and he specifies the parent with whom he was living. Leighton's different textual additions reveal his urge to revisit and re-narrate this image over time, even committing it to a recorded oral history.[88]

Leighton, who was ninety-four years old at the time he told his story for the image, was motivated to provide a personal narrative for the photograph for posterity, in a form that might persist after his death, and might allow his contribution to family history to be passed down to later generations.[89] Sharing these stories also allowed Leighton to quell anxieties about his inevitable death, which would come in 2006. By telling a story for these photographs, he was able to leave behind what he hoped would be a lasting record of his life, so that it might not be so quickly forgotten. As Batchen has suggested, adding signatures and other written notes to photographs is "a potent way to make a photograph more than a record of appearance," and "an unequivocal statement that 'I was there.'"[90] Indeed, this additional data further certifies his presence in the city, while explaining why he – a man who would live the majority of his life working and living on the family farm south of Quinter, Kansas – would have been photographed in St. Louis. For a span of two years, Leighton here reports that he was an urbanite, rather than always being a rural farmer. His stylish clothing suggests that he adapted readily, and perhaps even eagerly, to life in the city in the brief time he was in St. Louis with his father.

In this hand-written information, Leighton also mentions the angry disposition of the studio's dog, a comment that adds a touch of humor and vitality to this static conventional child's portrait. Would an animal that regularly was exposed to children as a "prop" at Raphael Studio be "cross," as Leighton suggests? While unlikely, it is possible. More probable, though, is the prospect that Leighton might have misremembered the demeanor of the dog or amalgamated it from memories of a number of dogs he knew from his youth. Our personal histories are admittedly formed, shaped, molded, fashioned, refashioned, enriched and invented by the malleable human memory – which, according to autobiographical writing expert Eakin, is a "source of fiction."[91] Misremembering, Eakin suggests, is a common condition of being human. Moreover, evoking the changeable associations of the human memory is one important function of family photographs.

Seizing the opportunity to narrate his own portrait further in an oral interview, Leighton eagerly offered these additional comments about his photograph:

> In 1914, we moved to St. Louis, where Dad worked for the streetcar company as a streetcar operator. I was around five years old and went to kindergarten school there. I enjoyed that very much. One day the class was taken to see the Liberty Bell, which was mounted on a railroad flat car and hung from a wooden frame. All of the kids went up and saw it and the teachers let us touch the bell. A few of them even kissed the bell. It really was the original Liberty Bell, *the* one with the big crack in its side. It was a very large bell. That was a big thrill.[92]

Curiously, the photograph reveals no trace of Leighton's father, streetcars, his schoolmates or the Liberty Bell. He does not mention the event of being photographed, or that his father had kidnapped him and took him to St. Louis because he disapproved of Leighton's mother working outside of the home as a teacher in a one-room schoolhouse less than a half-mile away from their Leighton homestead. In the oral history, Leighton does not speak of his father's eventual arrest, his parents' divorce, or his return to the family farmstead in rural western Kansas. Although Leighton surely missed the familiarity of the farm, his mother, his grandparents, and his friends in rural Kansas very much, he instead offers us an upbeat oral history to describe his experience in St. Louis.

Leighton's oral account does not contradict the image of the happy boy we see in the photograph with stories of what must have been a tumultuous and painful childhood. Studio portraits such as these, by convention, idealize moments of childhood for us to later revisit. Like the photograph, Leighton's oral history likewise offers only fond memories for posterity.[93] He saw this time in St. Louis as a rare adventure. Leighton does not feel sorry for himself for having a troubled childhood, through no fault of his own. Instead, according to his interview, Leighton made the most of his brief time in St. Louis – or so he claimed about ninety years later. Psychologists Michael Ross and Anne Wilson have discovered that the urge to "improve" one's past in the process of recounting personal histories is a nearly universal human tendency.[94]

One of the memories Leighton mentions specifically in his oral history involved what would be his only opportunity to see the Liberty Bell, as the artifact was on brief, well-publicized public display on a railroad car platform on its way back home to Philadelphia after being exhibited in the San Francisco Panama-Pacific Exposition. News reports confirm Leighton's account. The Liberty Bell stopped in St. Louis from 11:20 to 11:40 a.m. on Sunday, November 21, 1915, while on an extensive, slow-moving, cross-country rail tour (Fig. 3.6). Although the train carrying the bell stopped for just five minutes in smaller cities such as Greenville, Illinois, St. Louis was fortunate to have a twenty-minute window in which to celebrate the artifact's visit. At each stop, the Liberty Bell was met by a coordinated crowd cheering men, women and children, of bands playing patriotic music, soldiers making military salutes, and parades.

Leighton and his schoolmates participated in this fanfare, and seeing the Liberty Bell was one of the more memorable events of his childhood. He recalls the event of its visit with vivid detail. Leighton spoke of the bell's display on a wooden frame atop a flat rail car. He recalled stepping up onto the platform and touching the bell, and remarked that some people – excluding Leighton, who never liberally displayed physical affection – kissed the Liberty Bell. Archival photographs of the Liberty Bell's 1915 tour confirm these details about the bell's display and the interaction of visitors with the artifact.[95]

In 1915, many countries were engaged in World War I as America remained watchfully aloof. The bell's tour was designed, in part, as a publicity stunt to rally national unity, just in case America might need to cash in this patriotism and aid its allies by committing troops to the war. In the eyes of many Midwesterners, the Liberty Bell was associated with a remote East Coast culture and governing system that understood very little about what lie "out west." The bell's visit to the middle of America broke that region's sense of geographic isolation and gave Midwesterners a sense of connection to a history and identity beyond its own region. America's history, symbolized by the Liberty Bell, became *their* history, too.

As Leighton speaks of participating in this patriotic event, he connects to a master historical narrative much larger than just *his* story. Although Leighton's Midwestern

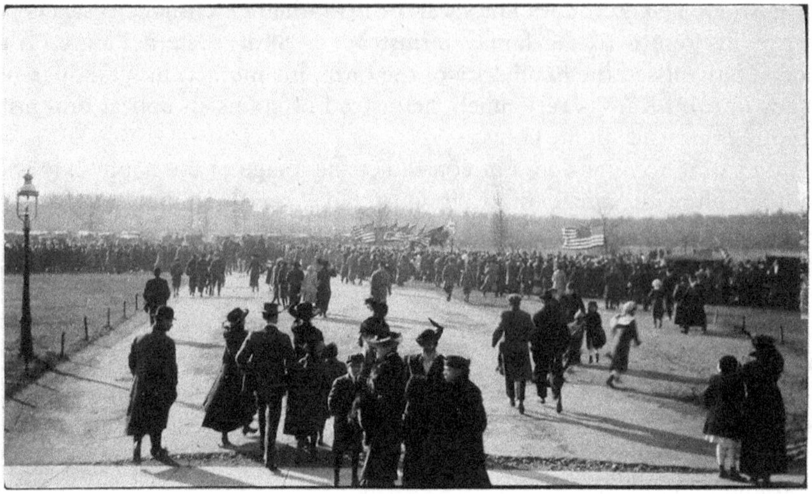

Figure 3.6 St. Louis Liberty Bell Visit, December 1, 1915 [537.295], The State Historical Society of Missouri, Photograph Collection. (Public Domain)

life may have seemed distant from that of America's founders, his relatives participated in the colonization of America and fought in just about every war in which the country was engaged.[96] Thus, the Liberty Bell's connection to the founding of America allowed Leighton to ponder his family's involvement in a broader temporal continuum, and the settlement of the country's frontier, encouraged by viewing and re-viewing the *Leighton Family* photograph and hearing its associated, hand-me-down narratives. Incidentally, Leighton is not unusual in this respect. Miller's urge to connect to a broader master narrative of the country's history, too, in the details she added to a poem that narrated her family's group portrait. In addition, Leighton's relationship to a broader historical master allows him to exceed the limited temporal scope of the human lifespan.

However, given its prominence in Leighton's oral history, the Liberty Bell's conspicuous absence from the image raises questions about the photograph's status as a prompt for the oral communication of memories that are only tangentially related to the subjects of the image. The photograph, instead, is a site for an amalgamated projection of "that-*never*-has-beens." As discussed previously, Olin and Knight have provocatively suggested that Barthes's "mistaken identifications" of objects in a photograph of his then-recently-deceased mother at age five suggest that viewers readily project *their own* narratives into photographs, regardless of the image's unique exposure-time-oriented narrative context.[97] As a photograph becomes a site for an amalgamated projection of various memories, it represents a genre of images so familiar that its physical presence may be regarded unnecessary, beyond being an initial springboard for storytelling.

The tendency for images such as Leighton's portrait to invite a wealth of changeable and subjective oral histories that do not correspond with objects or events visible in the photographs themselves is problematic for essentialist definitions of the image's function as an objective document. In many of our discussions of the medium – perhaps exemplified most clearly by the various commentaries offered in James Elkins's 2007 book *Photography Theory* – "indexicality," or "objective truthfulness" is considered in an absolutist claim that is "intertwined" with photography's "medium specificity," as Abigail Solomon-Godeau argues, and such claims have been contested in recent decades, in favor of a more nuanced view of the medium's functions.[98] The polemical nature of arguments about the "indexical" essence of the medium, Solomon-Godeau suggests, reflect various investments in the conceptualization of photography's history with respect to the continuing interest in Modernism's "medium specificity."[99] While Barthes wrote of photography's stilling of a "*that-has-been*," his "mistaken" identifications of items in photographs deconstruct any possible absolutism of this claim as he makes it. As such, Barthes implicitly argues that perhaps the medium's relationship to "the real" is variegated and complex. However, Kracauer also suggests that the human memory and photography operate in different realms, and adopts a binary conception of their functions:

> An individual retains memories because they are personally significant. Thus they are organized according to a principle that is essentially different from the organizing principle of photography. Photography grasps what is given as a spatial (or temporal) continuum; memory images retain what is given only insofar as it has significance. Since what is significant is not reducible to either merely spatial or merely temporal terms, memory images are at odds with photographic representation.[100]

As Kracauer suggests, memory and oral histories are inherently cognitively contained, and cannot possibly be encapsulated by, or bound to, the temporal/spatial matrix of the photograph – if indeed the medium's relationship to the real can be defined in such clear terms. Therefore, Kracauer's position suggests that Leighton's narrative has no obligation to coincide to what is visible in the photograph itself.[101] Family photographs such as Leighton's portrait instead invite the projection of oral stories that may take on the air of folklore and are not bound to the expectations of the documentary photograph.

Leighton's oral history also presents a challenge to the integrity of "objective histories," according to Nora. "Subjective" histories, he suggests, operate in opposition to "objective" histories because memory is selective, inherently biased, it perpetually evolves and it has a tendency to "deform reality."[102] Yet, insomuch as any recollection of the past is an active, constructive process – one that involves subjectivity, so that a historical narrative can be formed from bits of memory that were subjectively collected – even "objective" history is also revealed to be perhaps just as subjective as, for instance, Leighton's personal oral history. This argument, currently popular in memory discourse, theoretically (and perhaps idealistically) dissolves the notions of "objective" and "subjective" histories, and allows for the reworking of the very boundaries of history. It suggests possibility for the further exploration of oral histories and personal folklore – which often are dismissed in our photography-history discourses as frivolous. However, operating on the power of their ability to evoke personal history and memories, the uniquely subjective family photograph presents an opportunity for exploring the possibilities of a history that abandons the pretense of supposed "objectivity" in order to articulate the function of such images as objects that aid in the building and maintenance of personal memories and histories.

Leighton's oral history is necessarily shaped by the fallibility of human memory. His stories are partial, mutable, innately personal, and ripe with obvious omissions and the addition of absent items and/or events. It also is overly optimistic, revealing for posterity none of his parents' dramatic marital struggles or other unfortunate events that took a toll on him during the years in which this portrait was made. Leighton's oral history reveals that such accounts follow the same conventions that family portraits do: Representations of sad times or suffering are a violation of normative expectations. Instead, by convention, our family photographs and the stories they evoke should present an idealized view of a perfectly functional, happy family, jolly children, and lives lived free from pain, turmoil, or angst. They are nostalgic, fond, and light-hearted. Such histories may even refer to events and objects not even visible in the image itself. Family photographs such as Leighton's portrait instead invite the sharing of oral histories that often take on the air of folklore. But this is precisely why this early-twentieth-century genre of photographs was made, kept, and narrated. They operate in the realm of the subjective, in order to aid in the building and maintenance of personal memories and histories. Such stories, like the human experience itself, are amalgamations comprised of conflicting temporalities, built upon the surprisingly steady, yet essentially flawed, faculty of memory, tinged with optimism, nostalgia, and hope for the future.

Notes

1 Samuel Bernard Leighton identifies the subjects of this image as: "Front Row, L to R: Aunt Ella, Grandpa, Sam H. Leighton, Aunt Hazel, Grandma Cynthia Leighton, Uncle Sam D., Uncle Jacob, Back Row, L to R: Aunt Elybe, Grover My Dad, Aunt Lucendy, and Aunt Isa." His caption also gives the photograph the title *Leighton Family*.

2 Christina Kotchemidova, "Why We Say 'Cheese': Producing the Smile in Snapshot Photography," *Critical Studies in Media Communication*, Vol. 22, No. 1 (2005): 2–25.

3 Philip Stokes, "The Family Photograph Album: So Great a Cloud of Witness," in *The Portrait in Photography*, Graham Clarke, ed. (Seattle: University of Washington Press, 1992), 192–205.

4 Stokes, 192–205.

5 Betty J. Mills, "1889 Silk Taffeta Birthday Dress," *The Amanda Series: A Journal of Fashion History Through Paper Dolls* (Lubbock: Texas Tech University Press, 1991), n. p.; Jane A. Farrell, "Clothing for Adults in Iowa, 1850–1899," *Annals of Iowa*, Vol. 46, No. 2 (Fall 1981): 115, 119.

6 For more about how family snapshots relate to the projection of an image of social and economic success, see: Jeremy Seabrook, "Photography and Popular Consciousness: My Life Is in That Box," *Ten*, Vol. 8, No. 34 (Autumn 1989): 34–41.

7 Billie Hazel Leighton Flora, "Notes on Family History," Quinter, Kansas, Property of Karen (Leighton) Belden. Dover, Oklahoma (population: 367 in the 2000 census), is located in Kingfisher County, and is crossed by Highway 81. The Rock Island Railroad currently runs along the west side of this thirteen-street town.

8 Barthes, *Camera Lucida*, 13. Italics are my emphasis.

9 This poem was one of several Miller wrote, typed, and gave to Samuel Bernard Leighton. Miller wrote poems about many topics, including politics, the beauty of the frontier landscape, and growing up near Dover, Oklahoma. Her political poems capture a radical, leftist spirit whose loss is lamented in the recent popular novel *What's the*

Matter With Kansas? Thomas Frank, *What's the Matter With Kansas? How Conservatives Won the Heart of America* (New York: Metropolitan Books, 2004).

10 Liz Wells, *Photography: A Critical Introduction* (New York: Routledge, 1997), 150–151.

11 Geneaology.com, "Search for 'Samuel H. Leighton'" (accessed March 13, 2007); Perley Maynard Leighton, *A Leighton Genealogy: Descendants of Thomas Leighton of Dover, New Hampshire*, Vols. 1–2 (Boston: New England Historic Genealogical Society, 1989).

12 Stories about connections to these notable figures were part of the Leighton family's folklore. These ancestral links were confirmed by genealogical research.

13 Following the "Leighton" family name in genealogy records produced a lineage of shopkeepers and farmers, including Solomon Leighton (1739–1809), who served in the Revolutionary War, in New Hampshire's Second (Poor's) Regiment/Eighth Continental Regiment. Thomas Leighton, Jr. (1642–1677) served as a constable of Dover, New Hampshire. His parents, Thomas Leighton (1604–1672) and Joanna Silsby Leighton (1610–1703, married Thomas in 1634), came to the United States of America from Glasgow in 1633, aboard the fourth crossing of the *Mayflower*. This information was gleaned from an extensive search of records through the search engines at Ancestors.com and Genealogy.com, from June 2007 to July 2008. It should be noted that, while most Leighton descendants settled in the northern part of the country and fought for the Union in the Civil War, the first generation of Leightons to arrive in the United States kept a Native American man as a slave.

14 Author Unknown, *Kingfisher Panorama* (Kingfisher: Times Printing Co., 1957).

15 Michael H. Reggio, "Cheyenne-Arapaho Opening," *The Encyclopedia of Oklahoma History and Culture*, www.okhistory.org/publications/enc/entry.php?entry=CH031 (accessed March 13, 2018).

16 Reggio, n.p.

17 Reggio, n.p.

18 Reggio, n.p.

19 Reggio, n.p.

20 Samuel B. Leighton, "Interview With Author: Leighton Family History," Fall 2002.

21 Paul Ricoeur, "Narrative Time," in *On Narrative*, William J. T. Mitchell, ed. (Chicago: University of Chicago Press, 1981): 165–186.

22 Ricoeur, 165–186.

23 To read more about the railroad's role in provoking tensions between the individual and the globalizing aspirations of modernity's communications and transportation systems, see: Wolfgang Schivelbusch, *The Railway Journey: The Industrialization of Time and Space in the Nineteenth Century* (Berkeley: University of California Press, 2014), 42.

24 Nancy Ellen Leighton Miller, "Ok-Ar-Che" (Unpublished Poem), n.d. The Rock Island railroad line was placed on top of the Chisholm Trail in the 1890s: "A Brief Historical Overview of the Chicago, Rock Island, and Pacific Railroad," Rock Island Railroad.

25 Miller, "Ok-Ar-Che."

26 Miller, "Ok-Ar-Che." When they traveled by rail, the Leightons and Daltons would have encountered former slave, "Apostle" Paul Sykes, who came to Kingfisher in 1891, and began to preach on the streets of Kingfisher and in front of the jail. He also greeted passengers at the Kingfisher Depot by dancing a jig-step and singing the song "Old Ark's A-Movin'." After each performance, Sykes gathered coins thrown from the train. The lyrics to his song, which was a simultaneous ode to religion and to the new speed of transportation, are as follows:

If you expects to get to heaven when you dies
You must stop your tongue from tellin' lies
The old ark's a-movin', so move along, chillun.
The old ark's a-movin', movin' right along.
They called old Noah a foolish man
Because he built his ark on sandy land.
But the old ark's a-movin', so move along, chillun.
The old ark's a-movin', movin' right along.
I've never been to heaven, but I've been told

The streets up there are paved with gold.
And the old ark's a-movin', so move along, chillun.
The old ark's a-movin', movin' right along.
The author's spelling and grammar have not been changed.

Kingfisher Panorama (Kingfisher: Times Printing Co., 1957), n.p.

27 In the quote at the start of this chapter, Catherine Noren suggests that family snapshots can "turn time into a place" by bringing long-gone relatives into the present. Thus, she argues that the photograph is capable of collapsing both space and time: Noren, *The Way We Looked*, 5. The *Leighton Family* photograph also transports its viewers temporally into the Oklahoma territory, during the late frontier days, inverting the relationship Noren describes. This photograph and the narratives it evokes function as portals for enabling a viewer's imaginative, temporal presence in alternate times. However, as Roland Barthes points out, not all photographs inspire the same effect – or an effect at all – in every viewer. The *punctum*, for instance, is individually determined and subjectively variable: Barthes, *Camera Lucida*, 77.
28 Nancy Ellen Leighton Miller, "The Gumps's Pitcher Tooken" (Ellis, KS: n.d.).
29 Advertisement, "E. A. Carroll, Photographer," *Kingfisher Reformer*, August 31, 1894, Vol. 1, No. 6, 5.
30 Kris Belden-Adams, "Interview of Karen (Leighton) Belden" (Family History), September 12, 2007, New York, NY.
31 David Welky, "The Gumps: America's Comic-Strip Family," in *Everything Was Better in America: Print Culture in the Great Depression* (Bloomington: University of Illinois Press, 2007), 67–80.
32 Welky, "The Gumps," 69.
33 Welky, "The Gumps," 70.
34 Welky, "The Gumps," 70.
35 Welky, "The Gumps," 71.
36 Richard Chalfen, *Snapshot Version of Life* (Bowling Green, OH: Bowling Green University Popular Press, 1987).
37 Belden-Adams, "Interview of Karen (Leighton) Belden."
38 Kotchemidova, "Why We Say 'Cheese,'" 2–25.
39 Billie Hazel Leighton Flora, "Leighton Family Genealogy," Unpublished Records, 1985.
40 Pierre Nora, "Between Memory and History: Les Lieux de Mémoire," *Representations* 26 (1989): 8.
41 Michael Ross and Anne Wilson, "Constructing and Appraising Past Selves." In *Memory, Brain and Belief* (Cambridge: Harvard University Press, 2000), 233.
42 Miller's poem also fondly recalls a story about Samuel David, who lived in the same city as Miller around 1916. It is possible that the two may have had a stronger bond than she did with her other siblings.
43 Nora, "Between Memory and History," 8.
44 Kerwin Klein, "On the Emergence of Memory in Historical Discourse." *Representations*, No. 69 (Winter, 2000): 128.
45 Batchen, *Forget Me Not*.
46 Susan Pagé, "Interview," *Christian Boltanski – Compositions*, exhibition catalogue (Paris: ARC/Museé d'Art Moderne de la Ville de Paris, 1981), 7.
47 Flora, "Leighton Family History," n.d.
48 Pierre Nora, "Between Memory and History," 8.
49 John Paul Eakin, "Autobiography, Identity and the Fictions of Memory," in *Memory, Brain and Belief* (Cambridge: Harvard University Press, 2000), 290.
50 Siegfried Kracauer, "Photography," *Critical Inquiry*, Vol. 19, No. 3 (Spring 1993): 421–436.
51 To read more about "objective histories" and the binary positioning of history against memory, which Kerwin Klein argues is "one of the clichés of memory discourse," and is in great need of renovation; see Kerwin Klein, "On the Emergence of Memory in Historical Discourse," *Representations*, No. 69, Winter (2000): 128.

52 Cynthia Leighton would give birth to another boy four years later. He died during childbirth. Flora, "Leighton Family History," n.p.

53 Jeffrey Sconce, *Haunted Media: Electronic Presence From Telegraphy to Television* (Durham: Duke University Press, 2000), 157.

54 Sabine T. Kriebel, "Theories of Photography: A Short History," in *Photography Theory*, ed. James Elkins, *The Art Seminar* (New York: Routledge, 2007), 17–18; Barthes, *Camera Lucida*, 115–119.

55 Barthes, *Camera Lucida*, 96.

56 Doane, *The Emergence of Cinematic Time*, 4.

57 This idea is discussed further in: Barry Schwartz, "The Social Context of Commemoration: A Study in Collective Memory," *Social Forces*, Vol. 61, No. 2. (December 1982): 374–402.

58 Jean-François Lyotard argues that the act of narrativity itself is always driven by the viewer's *present*: "The narrative's reference may seem to belong to the past, but in reality it is always contemporaneous with the act of recitation." Jean-François Lyotard, *The Postmodern Condition* (Minneapolis: University of Minnesota Press, 1984), 22. In other words, Lyotard's conception of a photograph's observer-generated *narrative time* belongs to an eternal present. As this suspended present-tense enters temporal consciousness, however, it becomes the territory of the past and is readily replaced by subsequent associations. *Narrative time*, then, is fleeting and easily destabilized.

59 Samuel Bernard Leighton has a motivation for making reference to William Orman Leighton in the caption for the *Leighton Family* portrait. The two men were lifelong friends.

60 Flora, "Leighton Family History." This subtext is only accessible to viewers who have knowledge of the history behind the represented figures. Varying levels of viewer knowledge about an image's context affect the interpretation of photographs. As many photographs end up in antique stores, garage sales, flea markets, and even in galleries and museums, their narration is detached from the original family's folklore and gains the narrative associations projected upon them by complete strangers.

61 My memories of William Orman Leighton differ dramatically from his presence in Flora's inserted photograph. I recall meeting him when he was in his eighties, and nearing the endpoint of his "life cycle." This photograph provides for me (and surely, Flora, too) visual record of something I already know *logically* – that he once was a baby. But I do not know of this *experientially*. The photograph allows me to more easily imagine William's childhood.

62 Paul A. Harris, "Inventing," in *Time and Memory*, Jo Alyson Parker, Michael Crawford, and Paul A. Harris, eds. (Boston: Brill, 2006), 128.

63 Langford, *Suspended Conversations*.

64 Incidentally, Flora would not return letters requesting an interview about this photograph from 2005–2006, when she would have been in her mid-eighties. This alone might have been motivated by a concern about ceding narrative authority to another voice – or, worse yet, to nobody at all. White, Hayden. "The Value of Narrativity in the Representation of Reality," in *The Content of Form* (Baltimore and London: Johns Hopkins Press, 1978), 1–25.

65 Barthes, *Camera Lucida*, Part Two.

66 Victor Turner and Jacques Derrida separately argue that non-linear narratives express a degree of disorder which is a characteristic of our cultural context: Victor Turner, "Social Dramas and Stories About Them," *Critical Inquiry*, Vol. 7, No. 1 (Autumn, 1980): 141–168; Jacques Derrida, "The Law of Genre," *Critical Inquiry*, Vol. 7, No. 1 (Autumn, 1980): 55–81.

67 Walter Benjamin, "Little History of Photography" (1931), in *Walter Benjamin: Selected Writings, Vol. 2: 1927–1934*, Michael W. Jennings, Howard Eiland, and Gary Smith, eds., Rodney Livingstone, trans. (Cambridge, MA: Belknap Press of Harvard University Press, 1999), 510–511.

68 Carolin Duttlinger, "Imaginary Encounters: Walter Benjamin and the Aura of Photography," *Poetics Today*, Vol. 29, No. 1 (Spring, 2008): 84.

69 Duttlinger, "Imaginary Encounters," 79. She continues to suggest that "[t]his conception has far-reaching consequences not only for the theory of photography but also for its role within literature, as is suggested by Benjamin's empathetic engagement with the Kafka photograph and its incorporation into his own life story," 79. Duttlinger's recent scholarship explores Benjamin's role with literary studies, rather than his writing's impact upon photography-history discourses: Carolin Duttlinger, *Kafka and Photography* (Oxford: Oxford University Press, 2007). Interestingly, Duttlinger's use of the term "aura" often is synonymous with "temporality." She accommodates multiple auras, such as the alternative aura created by Benjamin's mistaken identification.

70 Margaret Olin, "Touching Photographs: Roland Barthes's 'Mistaken' Identification," *Representations*, Vol. 80 (Fall 2002): 114.

71 Roland Barthes, *Roland Barthes par Roland Barthes* (Paris: Seuil, 1975); Olin, "Touching Photographs," 114.

72 Olin, "Touching Photographs," 105–107; Barthes, *Camera Lucida*, 53, 87–88.

73 Olin, "Touching Photographs." She further argues: "Like the pearls that were exchanged for a slender ribbon, the distance between the description and the reproduction of the photograph may have disguised the resemblance for some, although here it is the photograph, rather than the description, that is delayed." Olin, "Touching Photographs," 109.

74 Olin provides an expanded definition of the *punctum*: "The *punctum* is the detail that is not there, or that one wishes were not there. Absence, in this book about loss, is presence. Like Jean-Paul Sartre's mental image in *L'Imaginaire*, to which *Camera Lucida* is dedicated, the punctum is 'a certain way an object has of being absent within its very presence,' or perhaps present within its absence." Olin, "Touching Photographs," 110; Jean-Paul Sartre, *L'Imaginaire: psychologie-phénoménologique de l'imagination* (Paris, 1940); translated as *The Psychology of Imagination* (New York: Philosophical Library, 1948), 104.

75 The *Leighton Family*'s *then-becoming-increasingly-symbolic* status became a reality. This manuscript of Miller's poems is the only known copy, and its previous owner merely glanced at it.

76 Knight, *Barthes and Utopia*, 265–266. As Olin explains, "few readers were willing to follow this twist of the plot." However, she admits that she was convinced: "It never seriously occurred to me that there really was a Winter Garden Photograph, but, like Knight, I have had mixed success convincing others." Olin, "Touching Photographs," 110.

77 Barthes, *Camera Lucida*, 67–69.

78 Olin, "Touching Photographs," 111.

79 Olin, "Touching Photographs," 108; Walter Benjamin, "A Liitle History of Photography," in *Classic Essays on Photography*, Alan Trachtenberg, ed., P. Patton, trans. (New Haven, CT, 1980), 206. Barthes writes that he chose not to reproduce the Winter Garden photograph in his book because it would mean nothing of significance to his readers. Barthes, *Camera Lucida*, 73.

80 Barthes, *Camera Lucida*, 72.

81 Olin, "Touching Photographs," 110–111. By attempting to know his then-deceased mother, Barthes enacts a hopeful project of resurrection that Ricoeur mentions: Ricoeur, "Narrative Time," 165–186.

82 Jan Campbell and Janet Harbord, eds., *Temporalities, Autobiography and Everyday Life* (Manchester, UK: Manchester University Press, 2002), 9.

83 Campbell and Harbord, *Temporalities, Autobiography and Everyday Life*, 8.

84 Barthes, *Camera Lucida*, 73.

85 *Material Memories*, a volume of essays edited by Marius Kwint, Christopher Breward, and Jeremy Aynsley, argues that a photograph's material form often is overlooked in favor of its content. Marius Kwint, Christopher Breward, and Jeremy Aynsley, eds. *Material Memories* (Oxford: Berg Press, 1999). This is not to deny the tactile, material potency of many photographic practices, including other forms of nineteenth-century portraiture that was meant to be held in the hand and engages a sense of touch or smell through the addition of flowers, human hair, etc. For an examination of the importance

of touch, see: Elizabeth Edwards and Janice Hart, eds., *Photographs Objects Histories: On the Materiality of Images* (London: Routledge, 2004).

86 Leighton commented in unrecorded conversations with his daughter Karen (Leighton) Belden that upon his return to rural Western Kansas, his schoolmates picked on him for dressing like an urbanite.

87 David A. Lossos, "Early St. Louis Photographers" [On-line Database, *Genealogy in St. Louis*], updated December 21, 2009. Available at: http://genealogyinstlouis.accessgenealogy.com/tracylewis.htm. (accessed February 30, 2010).

88 Samuel Bernard Leighton, Oral History Based on Three Family Photographs of His Selection, Spring 2004. Recorded by Jane Leighton Belden in Olathe, Kansas, on behalf of Kris Belden-Adams.

89 Langford, *Suspended Conversations.*

90 Batchen, *Forget Me Not*, 41.

91 Eakin, Paul John, "Autobiography, Identity and the Fictions of Memory," *Memory, Brain and Belief* (Cambridge, MA: Harvard University Press, 2000), 290.

92 Samuel Bernard Leighton, Oral History Project, Rural Quinter, Kansas, Fall 2003. Italics are intended to reflect his spoken emphasis.

93 For more about the subjects that are deemed appropriate in portraits and snapshots, see: Chalfen, *Snapshot Versions of Life.*

94 Ross and Wilson, "Constructing and Appraising Past Selves," 233.

95 Archival photographs of the Liberty Bell in St. Louis are in: Glass Plate Photography, Thomas Kempland Collection, J. R. Eike Photography, St. Louis, Missouri, 1914–1917.

96 Geneaology.com, search for "Samuel H. Leighton" (accessed March 13, 2007). Perley Maynard Leighton, *A Leighton Genealogy.*

97 Knight, *Barthes and Utopia*, 265–266; Olin, "Touching Photographs," 110.

98 Abigail Solomon-Godeau, "Ontology, Essences, and Photography's Aesthetics: Wringing the Goose's Neck One More Time," in *Photography Theory*, James Elkins, ed. (New York and London: Routledge, 2007), 265.

99 Solomon-Godeau, "Ontology, Essences, and Photography's Aesthetics," 259–269.

100 Kracauer, Siegfried, "Photography," *Critical Inquiry* 19 (1993): 47–63.

101 The same cannot be said for other photographic practices – including mug shots, passport photographs, driver's license pictures, forensic images, legal evidence, scientific photographs and photojournalism – which operate under the expectations of providing a combination of narrative and image that have an absolute truth value. The next chapter addresses the racing photo finish, which is held by a court of law as absolutely accurate to the order of race finished, for example.

102 Nora, "Between Memory and History," 8.

Bibliography

Armer, Michael and Schnaiberg, Allan. "Measuring Individual Modernity: A Near Myth." *American Sociological Review*, Vol. 37, No. 3 (1972): 301–316.

Arnheim, Rudolf. "The Two Authenticities of the Photographic Media." *Journal of Aesthetics and Art Criticism*, Vol. 51, No. 4 (1993): 537–540.

Baer, Ulrich. "The Shutter's Suspension of Time: Toward a Democritean Theory of Photography." In *Time Expanded*, Madrid: La Fábrica/PHotoEspaña, 2010. 43–51.

Baetens, Jan and Hilde van de Gelder, eds. *Photography and Time*. Leuven: Leuven University Press, 2010.

Baker, George. "Photography's Expanded Field." *October*, Vol. 114 (2005): 120–140.

Barthes, Roland. *Mythologies*. New York: Hill and Wang, 1972.

Barthes, Roland. *Image, Music, Text*. New York: Hill and Wang, 1977.

Barthes, Roland. *Camera Lucida: Reflections on Photography*. New York: Hill and Wang, 1981.

Batchen, Geoffrey. *Burning with Desire: The Conception of Photography*. Cambridge, MA: MIT Press, 1997.

Batchen, Geoffrey. *Each Wild Idea: Writing, Photography, History*. Cambridge: MIT Press, 2001.

Batchen, Geoffrey. *Forget Me Not: Photography and Remembrance*. New York: Princeton Architectural Press, 2006.

Batchen, Geoffrey. *Photography: Degree Zero: Reflections on Roland Barthes's Camera Lucida*. Cambridge, MA: MIT Press, 2009.

Batchen, Geoffrey and Heiferman, Marvin, eds. *Now Is Then: Snapshots from the Maresca Collection*. New York: Princeton Architectural Press, 2007.

Benjamin, Walter. "Little History of Photography (1931)." In *Walter Benjamin: Selected Writings, Vol. 2: 1927–1934*, Michael W. Jennings, Howard Eiland, and Gary Smith, eds. Rodney Livingstone, trans. Cambridge, MA: Belknap Press of Harvard University Press, 1999. 507–530.

Bergson, Henri. *Creative Evolution*. Arthur Mitchell, trans. New York: Henry Holt and Company, 1911.

Bergson, Henri. "Form and Becoming." In *The Philosophy of Time: A Collection of Essays*, Richard M. Gale, ed. Basingstoke, UK: Palgrave Macmillan, 1968. 397–405.

Bergson, Henri. "Time and Free Will: An Essay on the Immediate Data of Consciousness." In *Gilles Deleuze and the Ruin of Representation*, Dorothea E. Olkowski, ed. Berkeley: University of California Press, 1999. 121.

Borges, Jorge Luis. *Labyrinths: Selected Stories and Other Writings*, James Yates, ed. New York: New Directions Publishing, 1964.

Borradori, Giovanna. "The Temporalization of Difference: Reflections on Deleuze's Interpretation of Bergson." *Continental Philosophy Review*, Vol. 34 (2001): 1–20.

Bradshaw, C. M., ed. *Time and Behaviour: Psychological and Neurobehavioural Analyses*. London and New York: Elsevier Science B.V., 1997.

Brown, Marshall. "Literature in Time." *Modern Language Quarterly*, Vol. 65, No. 1 (2004): 1–5.

Brynteson, Susan and Melvin, L. Rebecca. *Self Works: Diaries, Scrapbooks and Other Autobiographical Efforts*. Newark, DE: University of Delaware, 1997.

Cadava, Eduardo. "The Image: A Monster of Time." In *Time Expanded*. Madrid: La Fábrica/PHotoEspaña, 2010. 27–41.

Campbell, Jan and Janet Harbord, eds. *Temporalities, Autobiography and Everyday Life*. Manchester, UK: Manchester University Press, 2002.

Cardwell, Sarah. "About Time: Theorizing Adaptation, Temporality and Tense." *Literature/Film Quarterly*, Vol. 31, No. 2 (2003): 82–92.

Carey, James W. *Communication as Culture: Essays on Media and Society*. Boston: Unwin Hyman, 1989.

Carr, Herbert Wildon. "'Time' and 'History' in Contemporary Philosophy; With Special Reference to Bergson and Croce." *Proceedings of the British Academy*, Vol. 8 (1921): 331–349.

Castells, Manuel. *The Rise of the Network Society*. Malden, MA: Blackwell, 1996.

Chalfen, Richard. *Snapshot Version of Life*. Bowling Green, OH: Bowling Green University Popular Press, 1987.

Chalfen, Richard and Hirsch, Julia. *Family Photographs: Content, Meaning, and Effect*. New York: Oxford University Press, 1981.

Crary, Jonathan. *Techniques of the Observer: On Vision and Modernity in the Nineteenth Century*. Cambridge, MA: MIT Press, 1990.

Crary, Jonathan. *Suspensions of Perception: Attention, Spectacle, and Modern Culture*. Cambridge: MIT Press, 1999.

Curie, Gregory. "Visible Traces: Documentary and the Contents of Photographs." *Journal of Aesthetics and Art Criticism*, Vol. 57, No. 3 (1999): 285–297.

Cutshaw, Stacey McCarroll, and Barrett, Ross. *In the Vernacular: Photography of the Everyday*. Seattle: University of Washington Press, 2008.

de Lauretis, Teresa, Huyssen, Andreas and Woodward, Kathleen, eds. *The Technological Imagination: Theories and Fictions*. Madison: Coda Press, Inc., 1980.

Deleuze, Gilles. "Mediators." In *Incorporations*, Jonathan Crary and Sanford Kwinter, eds. New York: Zone, 1992. 281–294.

de Man, Paul. "Roland Barthes and the Limits of Structuralism." *Yale French Studies*, Vol. 77 (1990): 177–190.

Derrida, Jacques. *Copy, Archive, Signature: A Conversation on Photography*. Stanford, CA: Stanford University Press, 2010.

Derrida, Jacques. *Memoires: For Paul de Man Reading*. New York: Columbia University Press, 1989.

Derrida, Jacques. *Of Grammatology (1976)*. Baltimore: Johns Hopkins University Press, 1981.

Derrida, Jacques. "Signature, Event, Context." *Glyph*, Vol. 1 (1977): 191–192.

Derrida, Jacques. "The Law of Genre." *Critical Inquiry*, Vol. 7, No. 1 (Autumn, 1980): 55–81.

Duttlinger, Carolin. "Imaginary Encounters: Walter Benjamin and the Aura of Photography." *Poetics Today*, Vol. 29, No. 1 (Spring 2008): 79–101.

Duttlinger, Carolin. *Kafka and Photography*. Oxford: Oxford University Press, 2007.

Eakin, Paul John. "Autobiography, Identity and the Fictions of Memory." In *Memory, Brain and Belief*, Daniel L. Schacter and Elaine Scarry, eds. Cambridge, MA: Harvard University Press, 2000. 290–306.

Edwards, Steve. *The Making of English Photography: Allegories*. University Park, PA: Pennsylvania State University Press, 2006.

Ellenberger, Henri. *The Discovery of the Unconscious: The History and Evolution of Dynamic Psychiatry*. New York: Basic Books, 1970.

Ellenbogen, Josh. "Photography and the Imperceptible: Bertillon, Galton, Marey." PhD diss., University of Chicago, 2005.

Ellenbogen, Josh. "Authority, Objectivity, Evidence: Scientific Photography in Victorian Britain." *Studies in the History and Philosophy of Science*, Vol. 39 (Spring 2008): 171–175.

Ellenbogen, Josh. "Camera and Mind." *Representations*, Vol. 101 (Winter 2008): 86–115.

Gehlen, Arnold. *Man in the Age of Technology*. New York: Columbia University Press, 1980.

Gere, Charlie. *Art, Time and Technology*. Oxford and New York: Berg, 2006.

Goldberg, Vicki. *The Power of Photography: How Photography Changed Our Lives*. New York: Abbeville Publishing Group, 1991.

Green-Lewis, Jennifer. "'Already the Past': The Backward Glance of Victorian Photography." *English Language Notes*, Vol. 44, No. 2 (2006): 25–43.

Griffiths, Alison. *Wondrous Difference: Cinema, Anthropology & Turn-of-the-century Visual Culture*. New York: Columbia University Press, 2002.

Grosz, Elizabeth, ed. *Becomings: Explorations in Time, Memory, and Futures*. Ithaca and London: Cornell University Press, 1999.

Hall, Edward Twitchell. *The Dance of Life: The Other Dimension of Time*. Garden City, NY: Anchor Press/Doubleday, 1984.

Halle, David. "Portraits: The Limitations of Likeness." *Art Journal*, Vol. 46, No. 3 (1987): 217–225.

Heidegger, Martin. *Being and Time (1927)*, Joan Stambaugh, trans. Albany: State University of New York Press, 1996.

House, John. "Seasons and Moments: Time and Nineteenth-Century Art." In *The Story of Time*, Kristen Lippincott, ed. London: Merrell Holberton, 1999. 194–197.

Humphreys, Kathryn. "Looking Backward: History, Nostalgia, and American Photography." *American Literary History*, Vol. 5, No. 4 (1993): 686–699.

Husserl, Edmund. *The Phenomenology of Internal Time-Consciousness*, Martin Heidegger, ed. Bloomington and London: Indiana University Press, 1966.

Husserl, Edmund. *The Crisis of the European Sciences and Transcendental Phenomenology: An Introduction to Phenomenological Philosophy*, David Carr, trans. Evanston: Northwestern University Press, 1970.

Huyseen, Andreas. *After the Great Divide: Modernism, Mass Culture, Postmodernism*. Bloomington and Indianapolis: Indiana University Press, 1986.

Huyseen, Andreas. *Twilight Memories: Marking Time in a Culture of Amnesia*. London and New York: Routledge, 1994.

Huyseen, Andreas. "Present Pasts: Media, Politics, Amnesia." *Public Culture*, Vol. 12, No. 1 (2000): 21–38.

Johnston, Adrian. *Time Driven: Metapsychology and the Splitting of the Drive*. Evanston, IL: Northwestern University Press, 2005.

Jonsson, Bodil. *Unwinding the Clock: Ten Thoughts on Our Relationship to Time*. New York: Harcourt, Inc., 1999.

Kern, Stephen. *The Culture of Time and Space, 1880–1918*. Cambridge: Harvard University Press, 1983.

Kern, Stephen. "Time and Art in Twentieth-Century Culture." In *Tempus Fugit: Time Flies*, Jan Schall, ed. Kansas City, MO: Nelson-Atkins Museum of Art, 2000. 22–34.

Kingfisher Panorama. Kingfisher, OK: Times Printing Co., 1957.

Klein, Kerwin. "On the Emergence of Memory in Historical Discourse." *Representations*, Vol. 69 (Winter 2000): 127–150.

Klinenberg, Eric. "Dying Alone: The Social Production of Urban Isolation." *Ethnography*, Vol. 2, No. 4 (2001): 501–531.

Knight, Diana. *Barthes and Utopia: Space, Travel, Writing*. Oxford: Clarendon, 1997.

Kotchemidova, Christina. "Why We Say 'Cheese': Producing the Smile in Snapshot Photography." *Critical Studies in Media Communication*, Vol. 22, No. 1 (2005): 2–25.

Kracauer, Siegfried. "Photography." *Critical Inquiry*, Vol. 19 (1993): 47–63.

Kracauer, Siegfried. *The Mass Ornament: Weimar Essays*. Cambridge, MA: Harvard University Press, 1995.

Kwint, Marius, Christopher Breward, and Jeremy Aynsley, eds. *Material Memories*. Oxford: Berg Press, 1999.

Langford, Martha. *Suspended Conversations: The Afterlife of Memory in Photographic Albums*. Montreal, London and Ithaca: McGill-Queen's University Press, 2001.

Legrady, George. "Image, Language and Belief in Synthesis." *Art Journal*, Vol. 49, No. 3 (1990): 266–271.

Leighton, Perley Maynard. *A Leighton Genealogy: Descendants of Thomas Leighton of Dover, New Hampshire*, Vols. 1–2. Boston: New England Historic Genealogical Society, 1989.

Levinson, Paul. *The Soft Edge: A Natural History and Future of the Information Revolution*. London and New York: Routledge, 1997.

Lienhard, John. *Inventing Modern: Growing Up With X-Rays, Skyscrapers, and Tailfins*. New York: Oxford University Press, 2003.

Lippincott, Kristen. *The Story of Time*. London: Merrell Holberton, 1999.

Lovejoy, Margot. "Art, Technology and Postmodernism: Paradigms, Parallels and Paradoxes." *Art Journal*, Vol. 49, No. 3 (1990): 257–265.

Lukacs, Georg. "Narrate or Describe?." In *Writer & Critic and Other Essays*, Arthur D. Kahn, ed. New York: Grosset and Dunlap, 1971. 110–148.

Marvin, Carolyn. *When Old Technologies Were New: Thinking About Electric Communication in the Late Nineteenth Century*. New York: Oxford University Press, 1988.

Marx, Leo. *The Machine in the Garden*. New York: Oxford University Press, 1964.

McElhone, John P. "The Signature of Light: Photo-Sensitive Materials in the Nineteenth Century." In *Beauty of Another Order: Photography in Science*, Ann Thomas, ed. New Haven: Yale University Press, 1997. 60–75.

McQuire, Scott. *Visions of Modernity: Representation, Memory, Time and Space in the Age of the Camera*. London: Sage Publications, 1998.

Miquel, Paul-Antoine. "Evolution of Consciousness and Evolution of Life." *MLN*, Vol. 120, No. 5 (2005): 1156–1167.

Morris, Wright. *Time Pieces: Photographs, Writing, and Memory*. New York: Aperture Foundation, Inc., 1999.

Nietzsche, Friedrich. *Human, All Too Human*. Gary Handwerk, trans. Stanford: Stanford University Press, 1999.

Nietzsche, Friedrich. *Friedrich Nietzsche*, O. Levy, ed. New York: Gordon Press, 1974.

Nora, Pierre. "Between Memory and History: Les Lieux de Mémoire." *Representations*, Vol. 26 (1989): 7–26.

Noren, Catherine. *The Way We Looked: The Meaning and Magic of Family Photographs*. New York: Lodestar Books, 1983.

Olin, Margaret. "Touching Photographs: Roland Barthes's 'Mistaken' Identification." *Representations*, Vol. 80 (Fall 2002): 99–118.

Olkowski, Dorothea E. "Time Lost, Instantaneity and the Image." *Parallax: A Journal of Metadiscursive Theory and Cultural Practices*, Vol. 9, No. 1 (2003): 28–38.

Ong, W. *Oralty and Literacy: The Technologising of the Word*. London: Methuen, 1982.

Orvell, Miles. "Reproduction and 'The Real Thing': The Anxiety of Realism in the Age of Photography." In *The Technological Imagination: Theories and Fictions*, Teresa de Lauretis, Huyssen, Andreas and Woodward, Kathleen, eds. Madison: Coda Press Inc., 1980. 49–64.

Orvell, Miles. *The Real Thing: Imitation and Authenticity in American Culture, 1880–1940*. Chapel Hill: University of North Carolina Press, 1989.

Osborne, Peter. "Photography in an Expanding Field: Distributive Unity and Dominant Form." In *Where Is the Photograph?*, Brighton: Photoforum and Photoworks, 2003. 47–62.

Pagé, Susan. "Interview." In *Christian Boltanski – Compositions*, exhibition catalogue. Paris: ARC/Museé d'Art Moderne de la Ville de Paris, 1981. 7.

Parker, Jo Alyson, Michael Crawford, and Paul Harris, eds. *Time and Memory*. Boston: Brill, 2006.

Peirce, Charles Sanders. *Collected Papers of Charles Sanders Peirce*, Charles Hartstone and Paul Weiss, eds. Cambridge: Harvard University Press, 1932.

Peirce, Charles Sanders. "Logic as Semiotic: The Theory of Signs." In *Semiotics: An Introductory Anthology*, Robert E. Innis, ed. Bloomington: Indiana University Press, 1985. 1–23.

Rabinbach, Anson. *The Human Motor: Energy, Fatigue, and the Origins of Modernity*. Berkeley: University of California Press, 1990.

Rabinbach, Anson. "Neurasthenia and Modernity." In *Incorporations*, Jonathan Crary and Sanford Kwinter, eds. New York: Zone, 1992. 178–189.

Ratliff, Floyd. "Illusions in Man and His Instruments." *Journal of Philosophy*, Vol. 68, No. 19 (1971): 591–597.

Reggio, Michael H. "Cheyenne-Arapaho Opening." *The Encyclopedia of Oklahoma History and Culture*, www.okhistory.org/publications/enc/entry.php?entry=CH031 (accessed March 13, 2018).

Ribiere, Mireille and Baetens, Jan, eds. *Time, Narration, and the Fixed Image*. Amsterdam and Atlanta: Rodopi, 2001.

Ricoeur, Paul. "Narrative Time." In *On Narrative*, William J. T. Mitchell, ed. Chicago: University of Chicago Press, 1981. 165–186.

Ross, Michael, and Wilson, Anne. "Constructing and Appraising Past Selves." In *Memory, Brain and Belief*. Cambridge, MA: Harvard University Press, 2000. 231–258.

Sandbye, Mettye. "Making Pictures Talk: The Re-Opening of a 'Dead City' Through Vernacular Photography as a Catalyst for the Performance of Memories." In *Re-Investing Authenticity: Tourism, Place and Emotions*, Britta Timm Knudsen and Anne Marit Waade, eds. Bristol, UK: Channel View Publications, 2010. 182–198.

Sartre, Jean-Paul. *L'Imaginaire: psychologie-phénoménologique de l'imagination*. Paris, 1940; Eng. trans. *The Psychology of Imagination*. New York: Philosophical Library, 1948.
Schall, Jan, ed. *Tempus Fugit: Time Flies*. Kansas City, MO: Nelson-Atkins Museum of Art, 2000.
Schivelbusch, Wolfgang. *The Railway Journey: The Industrialization of Time and Space in the Nineteenth Century*. Berkeley: University of California Press, 2014.
Schwartz, Barry. "The Social Context of Commemoration: A Study in Collective Memory." *Social Forces*, Vol. 61, No. 2 (December 1982): 374–402.
Schwartz, Sanford. *The Matrix of Modernism: Pound, Elliot, and Early Twentieth-Century Thought*. Princeton: Princeton University Press, 1985.
Sconce, Jeffrey. *Haunted Media: Electronic Presence from Telegraphy to Television*. Durham: Duke University Press, 2000.
Seabrook, Jeremy. "Photography and Popular Consciousness: My Life Is in That Box." *Ten*, Vol. 8, No. 34 (1989): 34–41.
Shaw, Marlow Alexander. "A Contribution to the Psychology of Time." *University of Toronto Studies: Psychological Series*, Vol. 1, No. 2 (1899): 101–153.
Shawcross, Nancy. "The Intertexts of La Chambre Claire: Barthes, the Photograph, and the Interstice of Time." Doctoral Dissertation. New Brunswick, N.J.: Comparative Literature, Rutgers, The State University of New Jersey, 1993.
Shawcross, Nancy. *Roland Barthes on Photography*. Gainesville: University of Florida, 1997.
Snyder, Carl. "The World Beyond Our Senses." *Harper's Monthly Magazine* (June 1903): 117–120.
Stokes, Philip. "The Family Photograph Album: So Great a Cloud of Witness." In *The Portrait in Photography*, Graham Clarke, ed. Seattle: University of Washington Press, 1992. 193–205.
Sturgis, Alexander. *Telling Time*. London: National Gallery Company Ltd., 2000.
Thomas, Sophie. "Making Visible: The Diorama, the Double and the (Gothic) Subject." In *Romantic Circles Praxis Series* (2005), www.rc.umd.edu/praxis/gothic/thomas/thomas
Turner, Victor. "Social Dramas and Stories About Them." *Critical Inquiry*, Vol. 7, No. 1 (Autumn 1980): 141–168.
Various Authors. "Vernacular Photography Special Issue," *Afterimage*, Vol. 19, No. 6 (May/June 2002).
Welky, David. "The Gumps: America's Comic-Strip Family." In *Everything Was Better in America: Print Culture in the Great Depression*. Champaign: University of Illinois Press, 2007. 67–80.
White, Hayden. "The Value of Narrativity in the Representation of Reality." In *The Content of Form*. Baltimore and London: Johns Hopkins Press, 1978. 1–25.
Wike, Lori. "Photographs and Signatures: Absence, Presence, and Temporality in Barthes and Derrida. *InVisible Culture*, No. 3 (2000): n.p.

4 Asynchronous, "Sculptural" Time and the Racing Photo Finish

A photo-finish image is the only legally binding, surefire way to determine the results of a race and the fate of millions of dollars each afternoon.[1] Just ask Don Morehouse, who has been taking photographs of horse-race finishes for more than forty-five years for the New York State Racing and Wagering Board:

> The photo finish has stood the test of time over the years. You can't get any truer than that. I've had guys ask me to fix race results. I tell them to come up and watch me photograph a race so they'll see for themselves that there's absolutely no way you can cheat this system.[2]

Reliable as they may be as evidence of the order of race winners or losers, photo-finish images are also full of visual distortions: horse's legs may disappear, rowers' oars and runners' legs often are elongated and appear to be made of rubber, the crew members who ply the oars are out of synchronicity with each other, and chariot and bike wheels are oval-shaped (Fig. 4.1).[3] What we see is illogical. *How could a chariot with an oval-shaped wheel move at all, let alone at the rate of fifty-five feet per second?*

Given these photo-finish subjects' departures from the appearance of our known reality, why then are photo-finish images held by the horse and dog-racing industries, the Tour de France, and the Olympic Games to be more trustworthy than the human eye at determining race placings? How do we reconcile their abstract nature with the fact that these images are deemed so reliable that they are binding in a court of law as the *only* fail-safe method to determine the fate of gold medals and millions of dollars? How can we have it both ways?

A photo finish does not behave as photographs conventionally do, with respect to time and space. It depicts a fixed vertical strip of space (the finish line) over a duration of time, using a moving strip of film that travels inside of a shutter-free camera at the estimated speed of the racing objects (Fig. 4.2). In the resulting still image, we see only objects that *were moving* as they passed a 4-inch-thick vertical strip of space at the finish line. The usual expectation that a photograph will reveal spatial relations at a given instant of time and in a full field of space – explored in Chapter 2 as a normative view of the medium – is reversed. Instead, the photo-finish image depicts a fixed vertical field of space, the finish line, over elapsed time. This is contrary to the way we are accustomed to relating to the passage of time, as it conventionally is represented in photographs, according to Dr. Andrew Davidhazy, who dedicated decades of his image-making career at the Rochester Institute of Technology to experimenting with variations on this image-making practice, "strip photography":

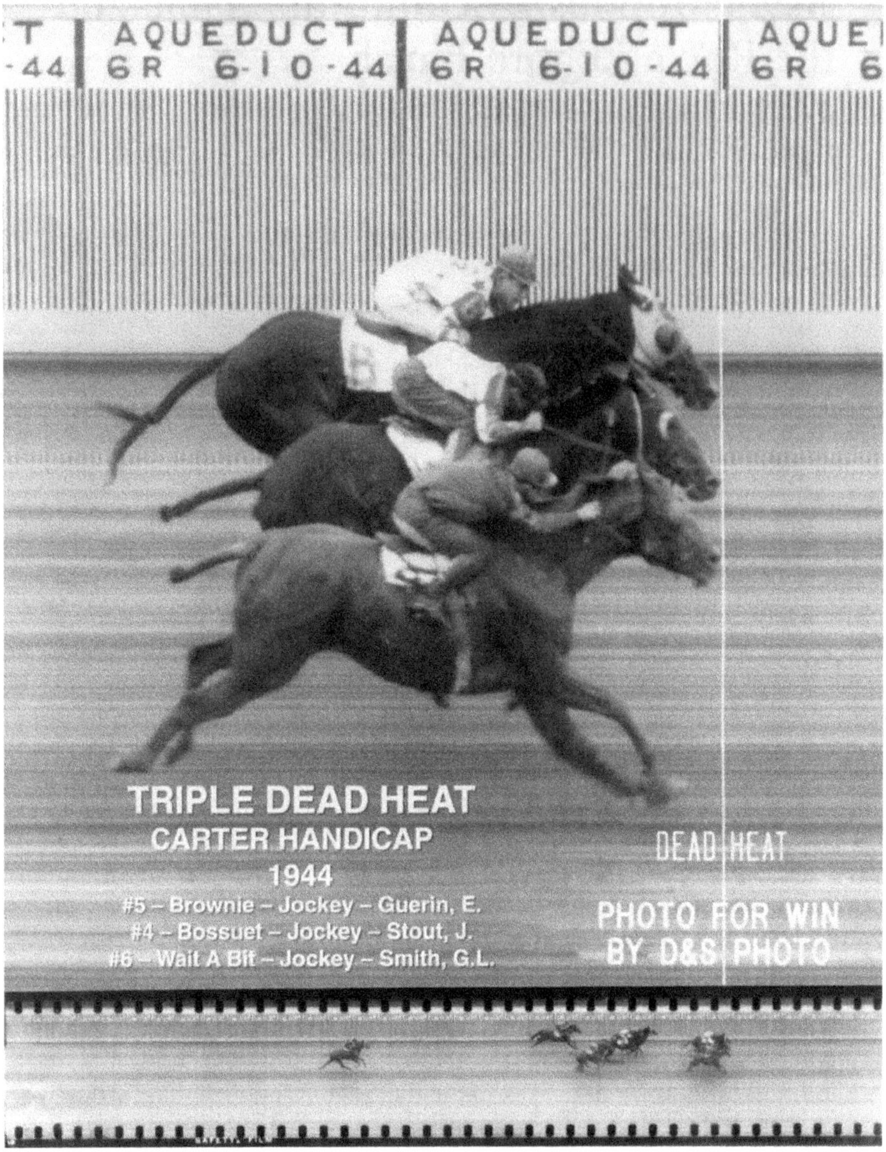

Figure 4.1 D&S Photo, *Photo Finish, Triple Dead Heat, Carter Handicap, Aqueduct Race Track, Jamaica, N.Y., June 10, 1944*, copy print, 8 in. x 10 in. (20.3 cm x 25.4 cm). Courtesy of Don Morehouse, D&S Photo.

> We are programmed to think about photographs as representing instants in time. And therefore, we have a horrible time relating to images that portray time itself as one of the dimensions of the final image … Most people don't think of time as the fourth dimension. It just *is*.[4]

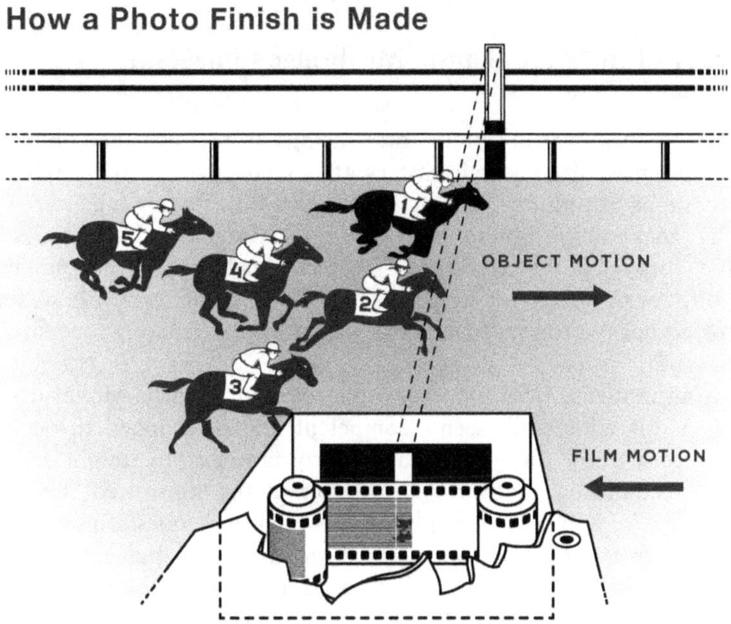

Figure 4.2 Tyler Barnes, *How a Photo-Finish/Strip Camera Works*, 2018. © T. Barnes Graphics.

As Davidhazy here suggests, due to the entrenched expectation that photographs provide "instantaneous" views of time, it itself is rarely employed as a variable, or a "sculptural" element of photographs, nor are time's considerations usually a precondition of comprehending the abstractions offered by a photograph. Photo finishes are made using cameras with modified focal planes, shutters, and/or film-advance systems. As a result, they offer a departure from the usual expectation that a photograph represents the appearance of a subject in a fixed, rectangular field of space at a fraction of a moment of time.[5] Instead, time and space in a photo-finish image is confoundingly asynchronous, accumulated, and distended, yet also compressed.

This chapter features a case study of the unique manipulations of time found in a vernacular and commonplace, yet understudied, application of strip photography, the racing photo finish, and its adjacent practices: peripheral (or "rollout") photography, the swing-lens panorama and Cirkut Camera photographs, and synchroballistic and aerial photography.[6] In addition to introducing readers to strip photography's unique reconceptualization of time and space, its early use and tendency to distort its subjects also sparked debates about the medium's position, vis-à-vis art and science. Through a close examination of the historical, social, and technical contexts of this photographic practice, this chapter will articulate the specific complications and possibilities that the sculptural representation of time in a photo-finish image presents for our current historical, cultural, scientific, and theoretical discourses.

A Pre-History of the Photo Finish: Muybridge's Images of "Instantaneous" Time

Before the photo finish was invented, horse-racing enthusiasts hailed photography for its perceived promise of temporal mastery after seeing Eadweard Muybridge's published photographs of running horses in 1878 (Fig. 4.3). Racing fans were specifically impressed by the medium's ability to provide a visual record of a fraction of a second of time that surpassed the 1/10th-of-a-second sensitivity of the human eye. An unnamed editor at the horse-racing periodical *The Spirit of the Times* was the first to suggest in print that photography might be useful for determining race finishes[7]:

> There is an immense field for electro-photography, which Muybridge, with the facilities at his command, seems competent and determined to occupy. It has occurred to us, by the way, that electro-photography might be turned to practical account on racehorses by depicting the finishes of heats. It seems entirely feasible, and would finally do away with all question as to the winner in a close finish. No suspicions could be cast upon a judge of this character, charges of bribery or partiality could not be made, and we commend the idea to Muybridge as one which may benefit him and the sporting public.[8]

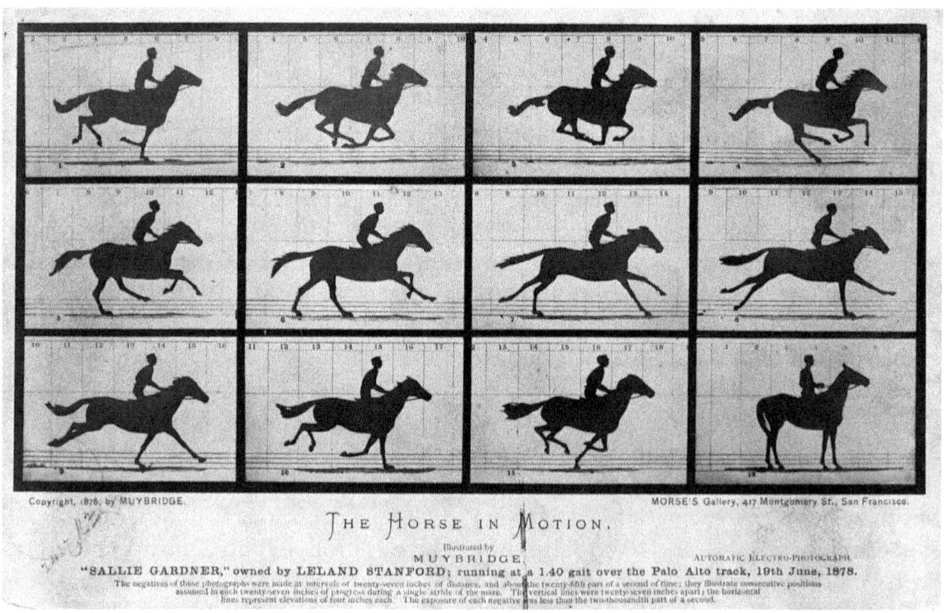

Figure 4.3 Eadweard Muybridge, *The Horse in Motion: Sallie Gardner, Owned by Leland Stanford, Running at a 1:40 Gait Over the Palo Alto Track, 19th June, 1878*. Public Domain.

Muybridge himself also claimed to have been the first to recognize the value of photography to determining the order of race finishes. In a letter published in 1882 – three years after *The Spirit of the Times* journalist's article was published – Muybridge wrote: "I venture to predict, in the near future that no race of any importance will be undertaken without the assistance of photography to determine the winner of what might otherwise be a so-called 'dead heat.'"[9]

Indeed, in horse racing and in historical writing, particular competitive emphasis is placed on *who crossed the finish line first*. However, to set aside debates about the origin of the idea to photograph race finishes for a moment, these comments – and a flurry of others sparked by them in various racing periodicals – collectively speak to a widespread enthusiasm for the fraction-of-a-moment-exposure photograph's potential to provide an unbiased visual record of the decisive span of time during which horses cross the finish line. Underlying that faith is the belief that an image made by a machine – the camera – has the capacity to master time and surpass the capacity of the unaided eye with unsurpassed objectivity. Cinema scholar André Bazin echoes this sentiment in his statement that photography is "an impassive mechanical process" that is therefore obligated to document the presence of objects in a specific temporal and spatial matrix:[10]

> The objective nature of photography confers on it a credibility absent from all other picture making. In spite of any objections our critical spirit may offer, we are forced to accept as real the existence of the object reproduced, actually *re*-presented, set before us, that is to say, in time and space.[11]

Placing faith in neutrality and infallibility of machines was widespread in the nineteenth and twentieth centuries. This subject has been much historicized by various scholars of technological modernity. Among them, historian Gerald Stanley Lee comments that the machine was regarded as "the symbol of the man's highest self. His own soul beckons to him out of it."[12] Lee also suggests that machines may transcend biological time and outlive their creators, provided the machine did not become obsolete in future generations.[13] As such, these objects might allow humanity to experience an infinite time that bypasses – and perhaps "cheats" – the biological and temporal limitations of the human lifespan.

Moreover, historian of media and technology Scott McQuire suggests that inventions such as the camera satiated positivist urges that prevailed from the time of photography's genesis – especially as the medium helped serve the needs of various fields of science:

> The machine was valued not simply because it was labour-saving, but because it was seen to surpass the capacities of the fallible human observer. Unstinting, methodical, uniform, durable, ever alert, incapable of being swayed by emotion or passion, the machine was invested as a paragon of Victorian virtues. Insulation from the spectre of individual judgment provided by all kinds of machines of measurement and recording consolidated the empirical orientation of the positivist project, in which physical appearances investigated by the armature of reason were increasingly treated as unambiguous facts.[14]

Technology historian Howard Besser even suggests that the grounded, unemotional machine might itself salvage objectivity from the destiny of dissolving into an anarchy of contingency.[15]

However, no matter how much stabilizing, empiricist solace Muybridge's photographs might bring, even those images represented departures from the appearance of objects in ongoing, continuous, lived time. They also were imperfect as visual proof of racing finishes, because the instantaneous photograph frequently missed the decisive finish-line-crossing moment. These problems can be further elucidated by taking a closer look at a set of Muybridge's photographs of horse Sallie Gardner, images which were broadly regarded as proof of the medium's temporal objectivity by the horse-racing community. So much so that editors of *The Blood Horse* magazine declared: "Anything written prior to instantaneous photograph was mere observation and opinion, and as such, can be discarded."[16]

Prior to the photographing of race finishes, Jerome O'Neill explains:

> The judging of horse races in this era was more casual than modern practice, and probably less accurate, as well. The judges stood in a low stand or platform, usually in the infield, on the opposite side of the track from the public. Thus, they neither had a particularly *good* view of the finish nor even the *same* view as the public. Such an arrangement was bound to lead to unpopular decisions, and in fact the judges' location may have been partly designed to give them some protection from outraged bettors![17]

Pre-History of the Photo Finish, Part Two: The Problems With the *Fraction-of-a-Second-Exposure* Image

In Muybridge's series of photographs titled *The Horse in Motion: "Sallie Gardner," Owned by Leland Stanford: Running at a 140 Gait Over the Palo Alto Track, 19th June, 1878*, a horse is depicted in silhouette, striding in front of a white gridded background marked by consecutively numbered segments of uniform size. Horse Sallie runs at an estimated speed of about 40 miles per hour past Muybridge's battery of twelve cameras, which each are equipped with stereoscopic lenses.[18] As the horse passes the cameras, she trips threads that release camera shutters.[19] The group of photographs represents a duration of elapsed time that is assumed to be linear and continuous, and is expressed in twelve separate images, each representing a light exposure of an estimated 1/1,000th of a second.[20]

As Muybridge's title implies, the series represents the narrative story of a horse in the act of running.[21] These images are "read" as a text might be in the Western world: in a linear sequence, from top to bottom, and from left to right. One-half-second-long gaps of temporal duration lie between each of Muybridge's images, implying elapsed unseen time, during which Sallie's body presumably moved to the position we see in the next frame.[22] Time's passage is therefore implied by changes in the spatial location of Sallie and her various body parts over time. These half-second-long durations of unseen time provide temporal spaces for viewers to fill with their own narratives, given the interpretive framework offered by Muybridge's descriptive title, *The Horse in Motion: "Sallie Gardner," Owned by Leland Stanford: Running at a 140 Gait Over the Palo Alto Track, 19th June, 1878.*

The series of photographs begins with Sallie in full gallop. Muybridge's title reinforces what we see by reminding us that the horse is *in motion*. Images and text collaboratively imply the existence of these represented moments within a temporal continuum, as part of an action that began *before* the 1/1,000th-of-a-second views of spatial relationships started being captured in the first photograph.[23] In other words, we do not see Sallie begin her run, but the images we do see could not have existed without those off-camera moments. But perhaps because Sallie's choppy start might not "read" as easily as a sequential series, Muybridge does not represent them, in favor of revealing the horse in full gallop, which provided closure to former California Governor Leland Stanford's bet with a friend about whether a running horse indeed has all of its hooves off of the ground simultaneously, in mid-run.

In defiance of Muybridge's title, Sallie's run ends amazingly abruptly in the final image (Fig. 4.4). In the twelfth photograph, numbers on the grid defy the previous sequential order, and Muybridge appears to have disregarded the estimated half-second interval that appears to have elapsed between the other frames. Even if the grid's numerical count repeated between the last two photographs, the rider would have undergone a sudden change in speed – from an average speed of covering *one* quadrant per half-second to a "slow-down" speed of covering at least *five* quadrants per half second. In other words, Sallie would have accelerated to five times her speed, to about 200 miles per hour, between the last two frames. This is implausible for two reasons: first, racing horses reach a running speed of about 55 miles per hour. Second, it would have been impossible for Sallie to come to an abrupt and complete stop in one-half of a second. The final image thus breaks the temporal cadence of a half-second gap of elapsed time between each 1/1,000th-of-a-second exposure, and it fractures the linear progression of time viewers are conditioned to expect by the other images in the grid.[24]

It is therefore likely that Muybridge added this image from another sequence of photographs in order to give an expected narrative closure to this series – and to indicate the completion of the act of running. Muybridge was known for making alterations to his other motion studies by adding non-sequential photographs to "complete" a set, as Marta Braun notes.[25] Muybridge's decision to insert a concluding image of Sallie at rest closes the series as a finished duration, as a "becoming" that finally "became." The final photograph in the series provides temporal closure, ending possible speculations about the fate of the run.

Although Muybridge's fraction-of-a-moment exposures offered visual departures from viewers' experiences of time as an ongoing continuum, the scientific community

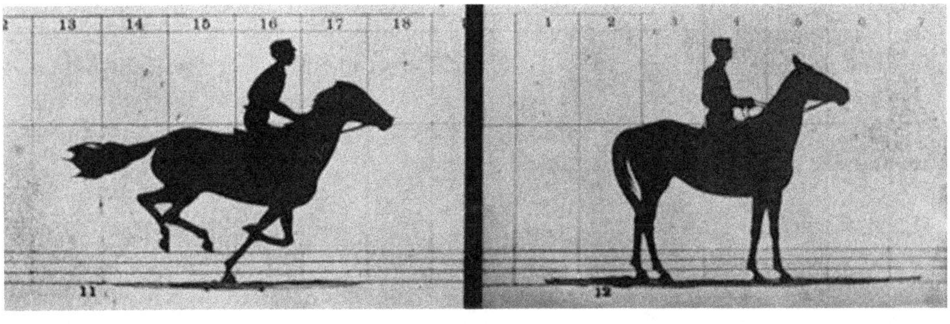

Figure 4.4 Eadweard Muybridge, Detail: *The Horse in Motion: Sallie Gardner, Owned by Leland Stanford, Running at a 1:40 Gait Over the Palo Alto Track, 19th June, 1878*. Public Domain.

readily accepted these photographs as "truthful" to the mechanics of equine movement.[26] In the decades prior to Muybridge's experiments, scientists granted photography the liberty of a fluid relationship to veristic realism when employed in service of the sciences. For instance, contact prints of botanical specimens made in the 1830s–1850s by William Henry Fox Talbot and Anna Atkins may have lacked surface detail and accuracy to the plants' specific hues, but these images still were regarded as persuasive and accurate documents. Microscopic photographs of Léon Foucault and Hippolyte Fizeau (c. 1845), daguerreotypes of the Moon made with the Great Refractor Telescope by George Bond and John Whipple (c. 1851), and Wilhelm Röntgen's X-rays (1895) all represented scientific realities inaccessible and unverifiable by the unaided human eye, but which nonetheless were eagerly accepted by the scientific community. Moreover, photography historian Josh Ellenbogen has argued that the forensic images of Alphonse Bertillon, the eugenics photographs of Francis Galton, and the motion studies of Étienne-Jules Marey similarly required a nuanced understanding of the operations of the photographic document.[27] This is to say, photography's ability to provide a *conditional* – rather than *absolute* – relationship to the real, for the sake of scientific illustration, was anticipated.

Muybridge's work, Marta Braun suggests, was motivated by his interest in mechanics, science, and by the imperative of uncovering the "laws that govern motion."[28] The photographs were largely successful in achieving that goal, she argues: "Although the resulting photographs showed hardly more than black silhouettes against a white background, they had an authority no drawing could ever claim."[29] Muybridge's images were regarded as purveyors of objectivity by the man who commissioned them.[30] In two of the frames of Muybridge's *Horse in Motion* series, all of Sallie's feet were off the ground. This provided the proof that Muybridge's patron, the tycoon/politician Leland Stanford, needed in order to win a bet.[31]

However, as Muybridge's "instantaneous" equine locomotion photographs were published in horse-racing-industry publications, they were regarded as perplexing, yet truthful, documents. An article written by the editor of horse-racing magazine *The Spirit of the Times* hailed Muybridge's photographs as "unerring," while also pointing out that "it is difficult, to the verge of impossibility, to explain why what we see with our own eyes on the race-course differs so much from what we see on the plate of the photographer."[32]

Art historian Thierry De Duve noted this difference and argued that Muybridge's instantaneous photographs represent a paradox in photography's representation of time: the medium's contradictory ability to represent with great apparent fidelity a series of instantaneous-time *movements*, while not representing *movement*, or corresponding to the appearance of subjects as they appear to us as we behold them in the ongoing continuum of lived time:

> Muybridge's snapshots of a galloping horse demonstrated what the animal's movements were, but did not convey the sensation of their motion. The artist must have felt squeezed between two incompatible truths that can be approached in terms of a contradiction in aesthetic ideology. But basically this contradiction is grounded in the paradoxical perception of photography in general, for which the example of Muybridge is simply an extreme case . . . The paradox of the unperformed movement and the impossible posture presents

> itself as an unresolved alternative. Either the photograph registers a singular event, or it makes the event from itself in the image. The problem with the first alternative is that reality is not made out of singular events; it is made out of the continuous happening of things.[33]

Jonathan Crary similarly remarks that the *Horse in Motion* series is a "blunt dismantling of apparent continuities of movement and time" in which "empirical similitude" is distanced from a "reality effect."[34] As a result, Muybridge's images, Crary argues, "exposed not naturalness of vision but … its systematic and constructed character."[35] Indeed, time in Muybridge's images is isolated into 1/1000th-of-a-second visual fragments, which are then reconstructed as a grid to evoke a narrative of duration. This series of photographs reveals not only the "constructed character" – to borrow Crary's term – of photography's representation of time, but of *time itself*. Both are abstractions.

This phenomenon did not go unnoticed by visual artists. The unique "reality effects" of instantaneous photographs such as Muybridge's sparked a heated discussion among artists, who debated whether the elegant horses in "flying gallop" poses painted by Théodore Géricault were superior to the awkward equines of Muybridge's photographs (Fig. 4.5).[36] French sculptor Auguste Rodin even thrust himself into the center of this debate, in a 1924 discussion with Paul Gsell about the instantaneous

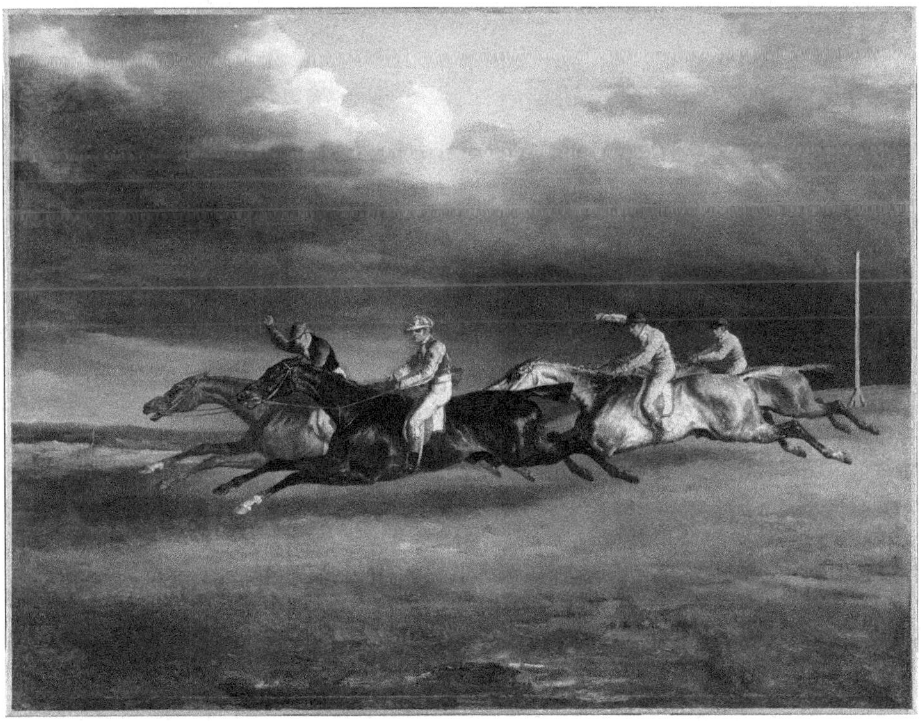

Figure 4.5 Théodore Géricault, *Derby/Racetrack at Epsom*, 1821. Oil in canvas, 92 cm x 122 cm. Musée du Louvre, Paris. © RMN-Grand Palais/Art Resource, NY.

photography of Muybridge and Marey, boldly declaring that "[i]t is the artist who is truthful, and it is the photograph that lies; for in reality time does not stop."[37] In his defense of painterly "truth," which he claimed was at odds with a photographic "truth," Rodin argues:

> Gericault is criticized because in his *Racetrack at Epsom*, from the Louvre, he painted horses that gallop ... in a familiar expression [the "flying gallop" position] ... [I]n the instantaneous photograph, the front legs of the horse are drawn in earlier, after having given propulsion to the whole body ... The four legs almost look as if they are suspended in air. They give the animal the appearance of sauntering in place, and of being immobilized in this position ...Gericault reasoned against the photographic [view of a moving horse] ... At the same time his view is false, it is truthful [, for instance,] when the start [of a gallop] is observed successively ..., since it is [Gericault's representation of the gallop] which we see, and that which impresses us.[38]

In Géricault's *Derby/Racetrack at Epsom*, all four horses have their legs extended outward from their bodies, as Rodin describes. This "flying gallop" posture, according to Irma B. Jaffe, was *the* conventional way of representing running horses in Western and Eastern art until the publication of Muybridge's book of animal locomotion studies in 1887 revealed this pose to be an "artistic convention" that was "not true to nature."[39] From that point on, the "flying gallop" disappeared from European art, Jaffe argues, and artists began representing horses in the climax of a gallop, with all four legs flexed under the body.[40] A new convention of representing running horses was based on emulating poses from Muybridge's instantaneous photographs of *The Horse in Motion.* In it, although the horse's legs are all off the ground at the same time in two instances (top row, middle two images), Sallie's legs are never extended in a "flying gallop." Rather, the horse contracts all legs, holding them close to her body. Even in the two frames of Muybridge's series in which all four legs are extended outward in a pose most similar to the one represented in Géricault's painting (middle row, last two images), one of Sallie's legs touches the ground. Muybridge's photographs suggested that horses do not ever position themselves – even for just a fraction of a moment – in a "flying gallop."

But, as Rodin argues, Géricault's preferred, painterly form of "truthfulness" was based on observations made with the unaided human eye – which does not see time as constituent, atomized units, but as a continuous flow of events and objects in motion. Rodin maintained that an artist committed to the "truthful" representation of nature should *only* depict what is visible to the eye, as it appears to the unaided human senses. According to Rodin, the photograph therefore "lies" at representing visual "realism."

Race-track patrons, on the other hand, also noted the difference between the "reality effect" of the optically centered visual experience and the abstract, camera-based one. However, they were dazzled by the camera's ability to surpass their own observations and operate by the logic of the temporal fragment. Racing fans regarded Muybridge's images as unbiased images of the mechanics of a horse's gait – even though the appearance of the animal in motion was less graceful than they were conditioned to anticipate.[41] In the eyes of racing fans,

Muybridge's images encapsulated photography's unique capacity to halt and capture a 1,000th of a second of time as a lasting visual document that may be viewed and reviewed in multiple points in later lived time.[42] This unique representation of photographic time, Bazin suggests, "embalms time, rescuing it simply from its proper corruption" – the "corruption" of our experience of time as a continuous, indivisible flow.[43] What was, to Rodin, a source of "truth" was to Bazin "truth's corruption."

However, cinema scholar Laurent Guido offers a different hypothesis. He suggested that serialized images by photographers such as Muybridge relayed an "obsession" with "speed characteristic of 'modern life,' including with visual solutions striving to represent movement in still form."[44] As such, Guido argues that such every photograph, in itself, encapsulates "the fleeting aspects of life."[45] Certainly, for patrons at a race track, the critical finish-line moment, and the passing of fortune that it brought, was all too fleeting, with potentially significant consequences for the bettor's future.

The Fraction-of-a-Second, Full-Field Photo Finish in Practice

The conventional full-field, fraction-of-a-second-exposure photograph was used to photograph race finishes starting in the late nineteenth century. Ernest Marx of Plainfield, New Jersey, claimed to have made the first single-exposure finish-line photographs in 1888, by panning his camera as horses passed the finish line and using a trip wire to release his shutters.[46] Published records can neither confirm nor deny whether Marx knew of Muybridge's prior use of trip wires in his Palo Alto experiments. However, because an article detailing Muybridge's experiments was published in one of the industry's most popular magazines, *The Spirit of the Times*, on June 3, 1882, and Muybridge's horse photographs first were published in *Scientific American* several years earlier on October 19, 1878, it *is* likely that Marx was familiar with Muybridge's techniques.[47] None of Marx's photographs survive today.[48] But race-track photographers, including Marx and J. C. Hemment – who began photographing races in 1890 – made sketches of their photo finishes for reproduction in racing periodicals. Two of Hemment's sketches of instantaneous photo finishes were reproduced on the June 26, 1886, front page of *The Spirit of the Times*. These sketches' existence suggests that the first fraction-of-a-second-exposure photo finishes may have been made two years earlier, in 1886.[49] The oldest surviving sample of this phase of photo-finish photography is Hemment's image of the Salvator-Tenny race from 1890 (Fig. 4.6).[50] Racing fans regarded these sketches as "documents" that testified to the truthful order of race finishes, even though they were produced by the human hand. The sketches' source material – photo finish photographs – lent the drawings credibility.

Horizontal Shutters, Moments Missed: Problems With the Single-Exposure Photo Finish

However, what once was a source of enthusiasm – the implementation of an unbiased machine, the camera, to produce a reliable testimony to the order of race finishes that exceeded the sensitivity of the human eye – was discovered to be fundamentally flawed. Almost immediately, problems were detected in the fraction-of-a-second-exposure photo finish. Almost all of the cameras employed on race

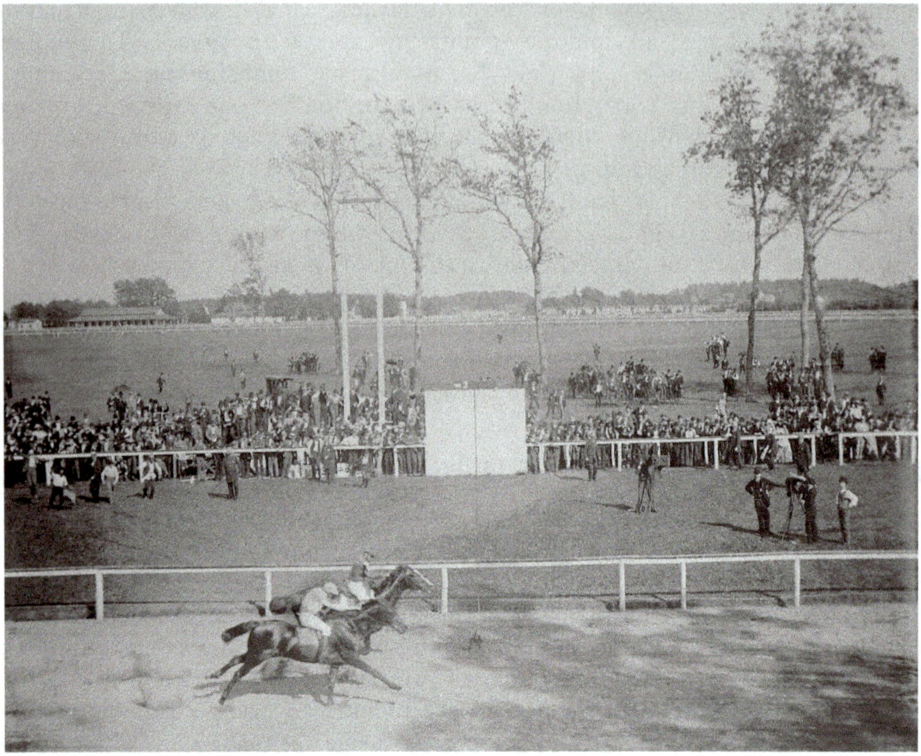

Figure 4.6 Keeneland Library, J. C. Hemment, Salvator-Tenny Match Race, Sheepshead Bay, NY, June 25, 1890. © Keeneland Library.

tracks used horizontally advancing camera shutters that exposed film from top to bottom, which allowed the horses closest to the photographer to be exposed on film slightly later. This bias – which advantaged the horses farthest from the camera – was discernible to the human eye. Once discovered, the horizontal-shutter advantage called the veracity of this photographic practice into question.[51]

Vertical shutters immediately were implemented as an industry standard, and this solution satisfied race-track patrons, owners, photo-finish technicians and journalists. Even with vertical shutters, the fraction-of-a-second-exposure photo finish was still deemed problematic. It frequently "missed" *the* decisive finish-line-crossing moment, and because it left subsequent finishes unrecorded.[52] In 1937 in France, a fraction-of-a-second-exposure photo finish actually incited a riot:

> At St. Cloud on July 4, placing judges, after looking at a photograph of the finish, declared Grande Beaute winner over Valombrosa, and backers of the latter set about smashing the judges' stand. So judges glanced again, awarded the race to Valombrosa. Backers of Grande Beaute smashed peri-mutuel booths, sent a police captain and two gendarmes to a hospital, about a dozen of their own number to jail. Mounted guardsmen restored order, eventually.[53]

The following colorful account of the same incident was offered by "Faraway," the French correspondent for London's *Horse and Hound*:

> Being anxious to hoist the photograph for public inspection, and allow racegoers to get away from the course, the film was put up, still wet and not completely developed … But as the photograph dried out it gradually appeared that the nose of La Valombrosa was in front of that of the grey, and naturally enough the mistake was immediately corrected by changing position of the numbers. This was the signal for one of the most disgraceful riots I have ever seen on a racecourse in any part of the world. … The crowd … broke windows, made bonfires of paper, attacked the peri-mutuel booths, and did much damage. The situation looked so ugly that, in addition to the large force of police, the mounted Garde Mobile has to be called out to take a hand, and it was not until an hour and a half had passed that the riot was quelled and the public left the course.[54]

Outrage and doubts about the credibility of the fraction-of-a-second photo finish peaked from the 1900s into the 1930s. United States sports journalist Ralph Powers writes of the general skepticism among race-track patrons in the 1930s (that led many bettors to be resourceful and bet on that day's fate of the cameraman himself):

> In some of the well known eastern tracks, the camera finish equipment was so poor that the average layman was willing to go back to the bookie, or bettor at the track, and make a second bet against the cameraman, so that he might cover his losses.[55]

As Powers notes, a loss of faith in the photo finish was directed at the faulty engineering of the camera mechanism and trip wires, not at the medium of photography's potential to isolate a decisive, finish line crossing, fraction of a moment of time. The hope of surpassing the human sensorium's capacity to process events was invested in photography – a medium that had proved that it could access unseen realisms. After all, despite the fact that the Great Depression caused race tracks to lose money, many chose to cut back on landscaping and did not hesitate to spend a staggering $300,000 in 1936 (the equivalent of US$5,396,744 in 2017) for Harry Day's new (soon-to-be-outdated), single-exposure camera, the "Eye in the Sky."[56] This decision paid off, because purses doubled when cameras were implemented at the most famous American race tracks – Santa Anita (California), Narragansett Park (Rhode Island), Arlington Park (Chicago, Illinois) and Hawthorne Park (Chicago, Illinois).[57] For instance, in Santa Anita, the "handle" – or, the total of all bets – on the day before the camera was installed was $221,537 (US$3,985,262 in 2017).[58] The next day it was $536,000 (US$9,642,183 in 2017). In just a few days, the cameras had paid for themselves.

The racing and mainstream media sang the praises of these new sensors, which patrons and reporters were certain had overcome the problems of the horizontal shutters and missed finish-line moments. Sportswriter Robert Harron of *Collier's Illustrated Weekly* wrote that: "this electric camera has reduced horse-race finishes to a scientific certainty, doing away with the error of oblique viewing and all other pitfalls." He continued his praise of the new technology:

> The announcement [of the race finish] was greeted with cheers, the purse was split, the jockey fees were split, and the customers were satisfied. Horse racing is one business-sports enterprise in which the customer can't always be right. If he were, somebody would go broke a great deal faster than you and I do now. But if we have to lose our money – and it was Damon Runyon who observed laconically that "all horse players must die broke" – I think it is only right that we lose it in the most painless way possible ... The Eye in the Sky has eased the pain in the vicinity of the pocket in which you carry your wallet, for you must admit that the shock of losing isn't nearly so painful when it is quick and beyond all doubt.[59]

Despite the race-track cameras' previous technical quirks, the public's faith in *photography* apparently never wavered, and it accommodated any necessary mechanical retooling. Even the praised Eye in the Sky system had its share of missed finish-line moments.

Engineers and photography technicians focused in the early decades of the twentieth century upon creating a camera that had the sensitivity and accuracy of the fraction-of-a-second exposure, but could more reliably capture the decisive finish-line-crossing moment. Engineers tried employing "photo cells" to trigger the release of faster camera shutters, high-speed shutters (the technology employed by Day's Eye in the Sky, as well as Gustavus Kirby's Two-Eye), and also experimented with movie cameras to increase the odds of capturing *the* finish-line-passage duration. The resulting images still often missed the critical moment of the race finish and were deemed unreliable. The racing industry's various engineers and technicians turned their attention to inventing a camera that never "blinked" and could record the finish line continuously, in the hope of capturing the crucial finish-line-crossing moments for not only the first racing horse, but later horses, too.

The Strip-Photography Photo Finish: A Camera That Did Not "Blink"

Multiple inventors from different countries staked the claim of being the first to create the strip-photography-based photo finish. British inventor Robert Capon filed for a patent in 1926 (Fig. 4.7) for a camera that, according to *The British Racehorse*, "is in fact the sire of the modern racecourse camera," even though it "does not appear to have been regarded by the inventor as having any particular application to horse racing."[60] In other words, it was a technological antecedent – even if it is not an antecedent in practical application. The Capon Camera was a shutterless camera that featured a vertical slit for an aperture and film that – if calibrated properly – moved at the same speed as the moving subjects and successively recorded moving objects over durations of time. The Capon Camera was used to photograph aircraft and automobiles as they were moving, and never was employed at a race track.[61]

Strip photography first was adapted specifically for race-track use in 1937 by Lorenzo Del Riccio, an optical engineer for Paramount Pictures' technical research laboratories, for debut at the inaugural horse race at Bing Crosby's Del Mar Turf Club – the luxurious race track of the rich and famous of Hollywood (Figs. 4.8 and 4.9). This technology became the standard in the Americas and in Europe. Although Bing Crosby is most well-known for being a singer, he also was a technophile. In addition to facilitating the adaptation of strip-camera technology for use at horse-racing tracks, he

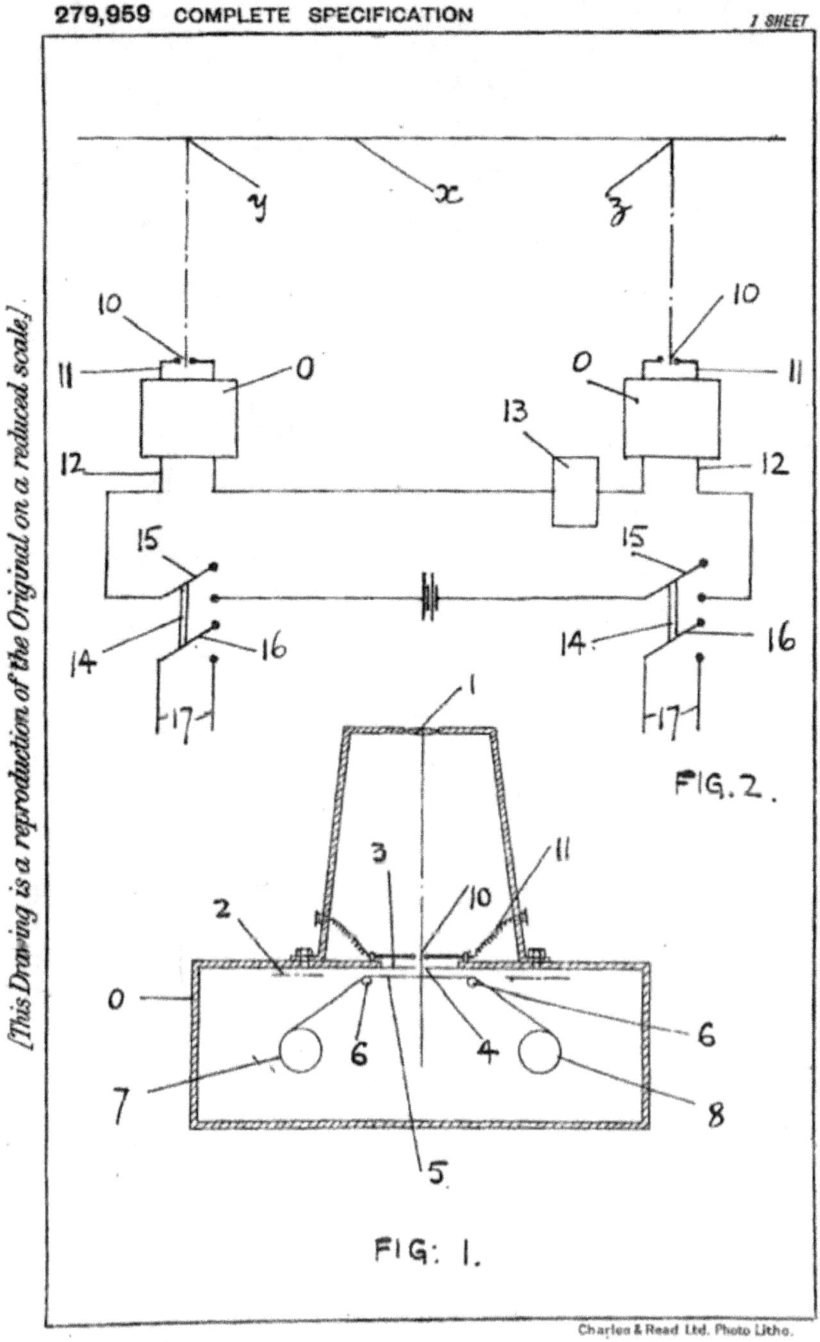

Figure 4.7 Robert Capon, Capon Camera, Patent No. 279,959. © Her Majesty's Stationery Office, British Patents.

June 1, 1943. L. DEL. RICCIO 2,320,350

CAMERA

Filed March 20, 1939 4 Sheets-Sheet 3

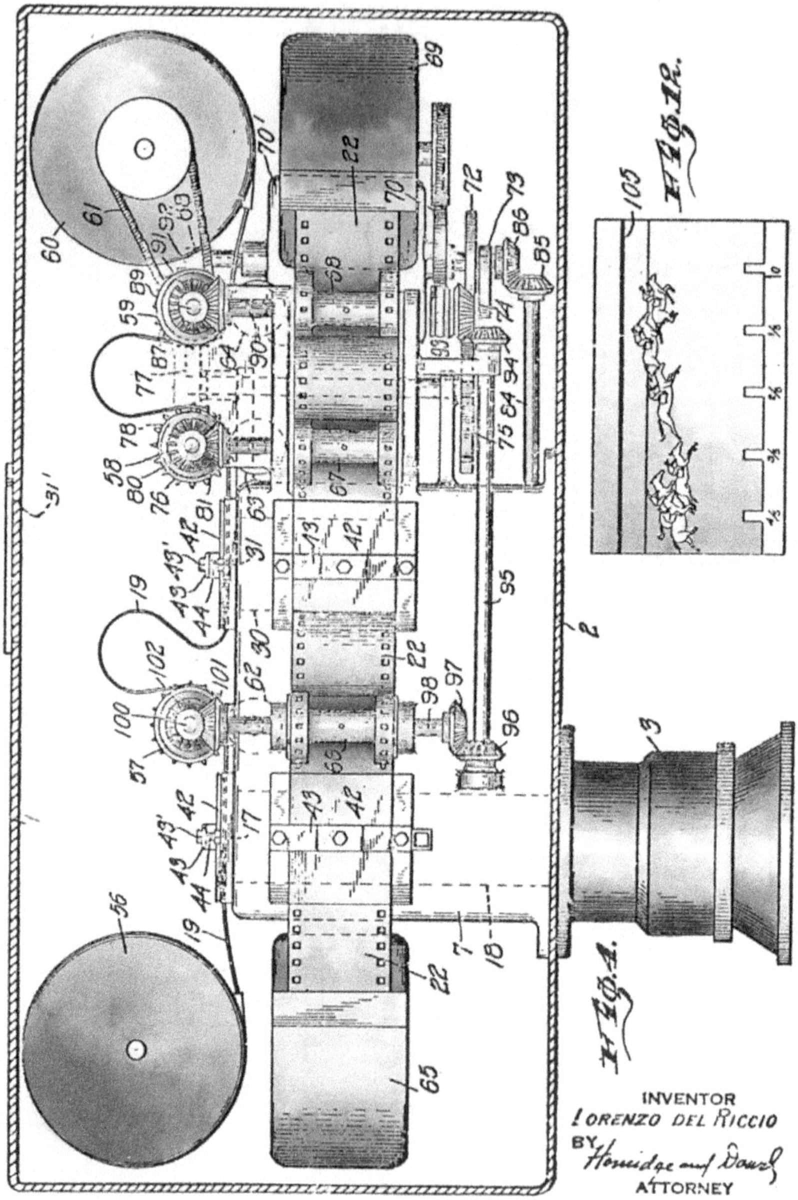

Figure 4.8 Lorenzo Del Riccio, *Patent Drawings*, Page 3 of 4, Patent No. 2,320,350, Issued June 1, 1943. © United States Patent and Trademark Office.

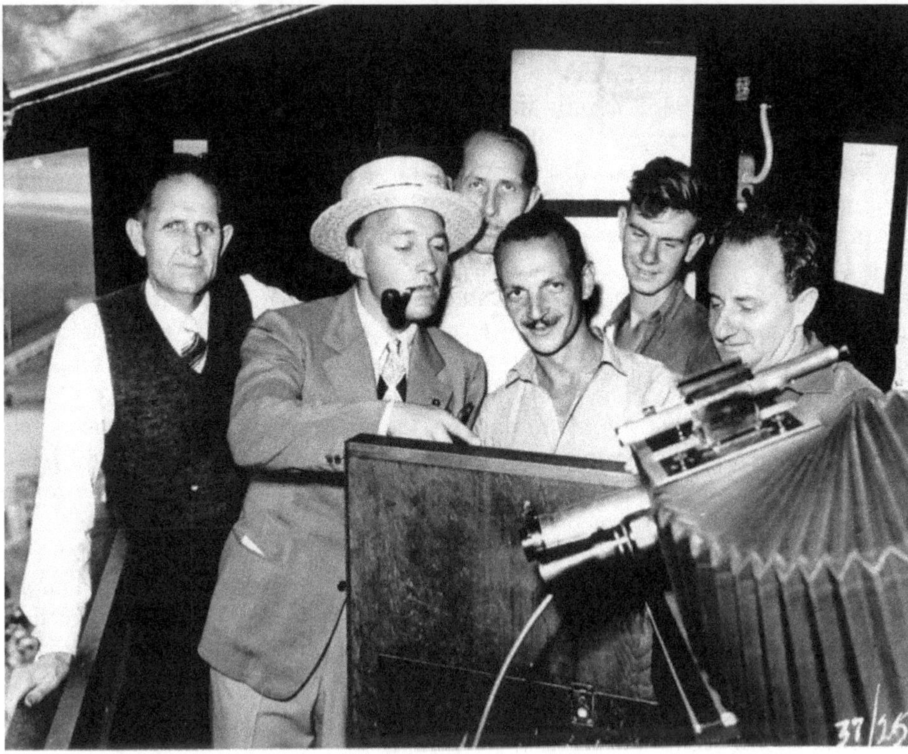

Figure 4.9 Lorenzo Del Riccio (with moustache) stands with Bing Crosby (in hat) and other managers of the Del Mar Turf Club next to Del Riccio's new photo-finish strip camera. The camera was first introduced at the opening of the Turf Club in 1937. The same technology is still used in the racing industry today. (Courtesy of Mac McBride, Del Mar Turf Club).

also pioneered the use of the microphone, and was instrumental in bringing the technology of magnetic tape recording from Germany to America. This new advance allowed entertainers to revise and perfect their radio performances, and even record them in advance. It also served to collapse the distinction between live performances and previously taped ones. Shows with tape-recorded segments allowed present and past temporalities to mingle freely, sometimes misleading listeners into thinking that the entire radio show was a live broadcast. Like photography, the invention of tape-recorded performances presented listeners with a baffling absence/presence conundrum.[62]

It should be noted that staking claim to inventing the photo finish, like the race to invent photography itself, also was a matter of national pride. After the Del Riccio camera had become popular in England, the British were swift to find grounds to contest Del Riccio's status as the creator of that technology. (Curiously, British journalists overlooked the British Museum's peripheral camera – the strip camera's ancestor technology. It will be discussed in detail shortly.)

Competing inventors and Australians Athol Shmith and B. Altson Pearl also laid claim to the invention of a strip camera for use to determine race finishes. The Shmith and Pearl camera system, or Camera-Graph, was based on the same technology as Del Riccio's and Capon's cameras. Even though their camera system only dated back to 1946, Shmith and Pearl added a mirror on the far side of the finish line to allow cameras to be able to "see" blocked horses and saddle numbers. This system became the standard in Australia, New Zealand, Africa, and Asia.[63] Within a few years, these mirrors soon became a global racing standard.

With the exception of upgrades such as the addition of timing systems, mirrors, and digital data processing to photo-finish practices, the basic camera technology has changed very little in the eight decades since its invention. Digital photo-finish cameras have been in widespread use since the late 1990s.[64] They have maintained the same aperture, focal-plane, and mobile-recording design as their film-based predecessors.

Australia, incidentally, is home to the only known strip-photography photo finish that *ever* has been publicly retracted as "untruthful." In a close finish, the 80:1-odds horse Rimfire was declared the winner over the more-favored Dark Marne for the 1948 Melbourne Cup.[65] It was the first time that photo finish equipment was used at the Flemington race track, and it was installed incorrectly. This mistake was discovered months later. The decision remained in the record books, and money was not refunded. Despite this error, strip-photography-based photo-finish technology enjoyed a warm reception in Australia, as well as in other countries worldwide.

The Time-and-Space Matrix of the Strip-Photography Photo Finish

To record a horse race, strip-photography photo-finish cameras are positioned in a booth in the grandstands above the finish line, and are focused on a 4- to 8-inch-thick vertical slice of space at the line (Fig. 4.2).[66] When horses approach the finish line, a technician activates the camera's film-advance motor by flipping a switch. In an analog (pre-digital) photo finish, film begins to move past the vertical slit of an aperture and is exposed to light. Racing horses pass the exposure space at (ideally) the pre-calibrated speed of the film's movement, and are recorded on film.[67] Only moving objects are recorded in strip photography. Static subjects appear as streaks, reversing both the conventional expectation that blurs are visual notations of moving subjects and the convention of anticipating that a photograph represents still objects within a full visual field.[68]

While a strip-photography photo finish might appear at first glance to offer a fraction-of-a-second view of objects within a full field of space, it instead provides a peculiar representation of time as a finite, relative duration. For instance, in the analog photo finish titled *Photo for Win, June 29, 1956, Roosevelt Raceway* (Fig. 4.10), six harness-racing horses finish the fifth race ofthe day. The image is rife with abstractions. The strip of ground at the finish line area remained still, so the turf is recorded on the film as a streaked field of grey. Only moving objects are recorded. Static objects become streaks. If racers' feet are planted at the finish line, they would be still, momentarily – and thus would not appear in the photo finish. This creates the "ghost leg" effect. In the photo finish, we see an abstraction, an elapsed "smear" of a brief, subjective span of accumulated durée that marks a departure from the manner that a full-field, fraction-of-a-second-exposure

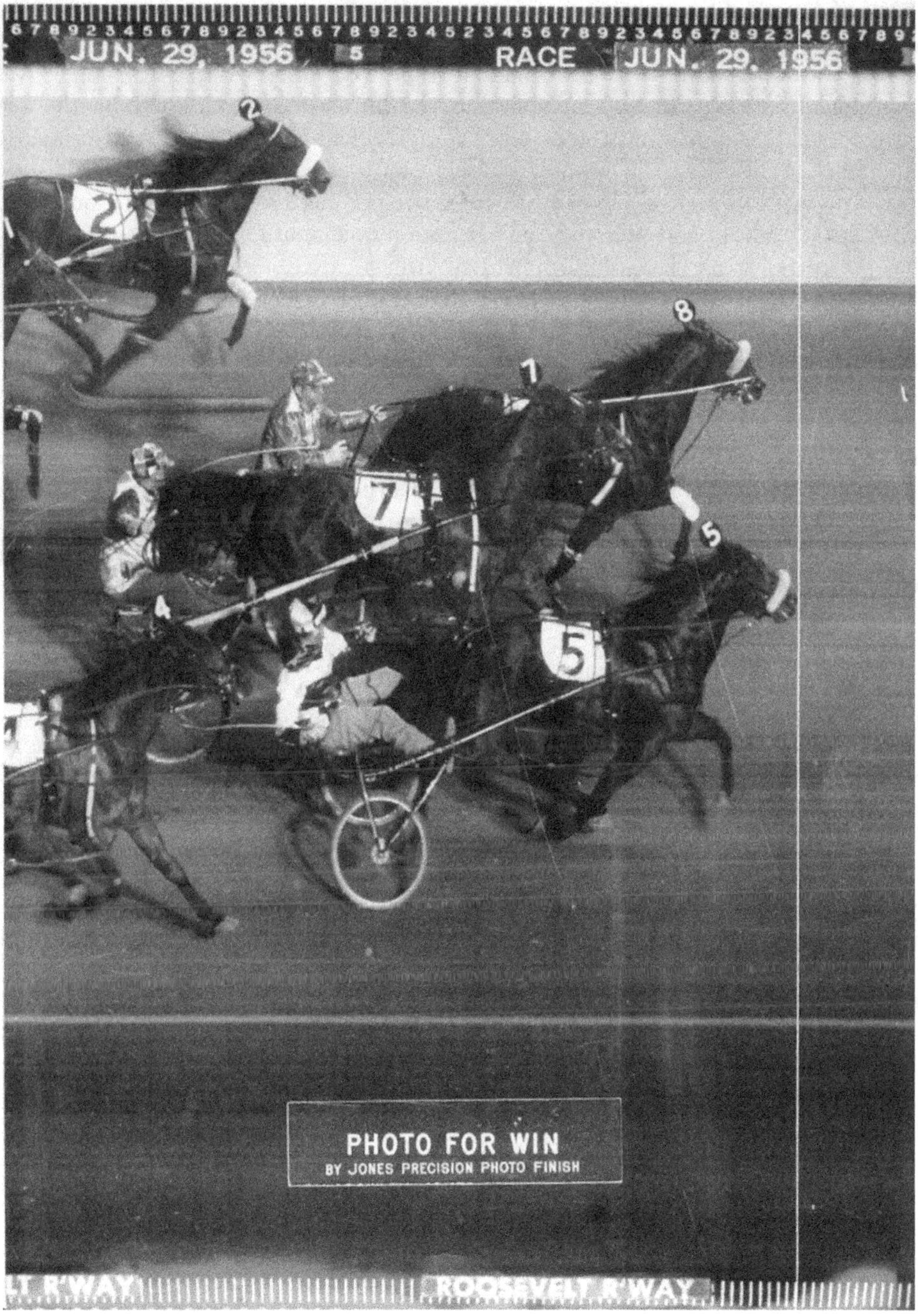

Figure 4.10 Unknown Photographer, Jones Precision Photo Finish, *Photo for Win, June 29, 1956, Roosevelt Raceway, Race No. 5*, 1956. Courtesy of Don Morehouse.

photograph relates to the time-and-space matrix.[69] The photo finish reveals a static image of a state of linear becoming, a state which ended with the achievement of the moment we see – the crossing of the finish line. This moment has great consequences in later time, including lost or gained fortune and glory. A span of subjectively determined time in the slice of space in the finish line is anointed by the very existence of the photo finish as a *more important* time and space than others we experience. That linear, finish-line-crossing, span of a few seconds of time is so very crucial monetarily that a camera has been deemed necessary to extend our visual sensitivity threshold and to boost our perception of spatial relationships over time in that prioritized space. In a photo finish, all "times" are not created equal, especially those whose consequences involve the exchange large sums of money.

Additional abstractions are apparent in *Photo for Win, June 29, 1956*. A crisp white "finish" line appears at the nose of the winning horse. It does not, however, bow in the wind, or adhere to the nose of the winning horse as it passes. In fact, racing patrons never would have seen this white finish line at all. Because the entire photograph represents an ongoing series of finish-line-crossing moments, *every* part of a photo finish represents the finish line. In order to produce the expected finish "line," and to provide a point for comparing the distance between one finishing horse's nose and another's, a photo-finish technician carefully placed a straight wire while in the darkroom in a perpendicular position to the track (using the short white vertical marks on the top and bottom of the image as guides), and exposed a piece of photographic paper with the wire on top (Figs. 4.11 and 4.12).[70] The wire obscures the thin strip of paper under it from being exposed to light. An analog photo-finish image is part contact-, part enlarger-exposed print. The edges of the image are also marked by the camera with notations that likewise were never seen by race-track patrons, including the date of the image, the name of the race track, sequential numbers from 2 to 9, and evenly spaced white marks. This information is imprinted on the film as it advances through a spindle inside of the camera.

In the *Photo for Win*, the horses' legs disappear and chariot wheels appear to be oval-shaped. The photograph fails to depict objects as they appear to us with our own vision. As noted in the introduction of this chapter, what we see is illogical. A chariot with an oval-shaped wheel could not move at all, let alone at the rate of 55 feet per second.[71] Given its many departures from reality and photographic convention, how do we reconcile the photo finish's abstract nature with the fact that it is so reliable and wholeheartedly trusted as the *only* fail-safe method to determine the winner of a race – and, thus, to determine the fate of gold medals or millions of dollars?

If the film in a photo-finish camera is calibrated to move faster or slower than the racers, then subjects will be distorted. Quicker objects appear compressed. Slower ones are elongated (Fig. 4.13). Wheels moving faster than 55 miles per hour (such as the ones on the harness-racing chariots in *Photo for Win*) appear oval-shaped. If a horse steps on the finish line as it passes the aperture's field of view, the horse's partially still leg will appear as a faint smudge on film, because only moving objects are recorded.[72]

Even more pronounced distortions occur in finish photographs for sports that involve racers with a number of moving parts or limbs. The photo finish is used to document race finishes in the sports of cycling, canoeing, motorboating, speed-skating, swimming, roller-skating, track events, dog-racing, and rowing. Rowing

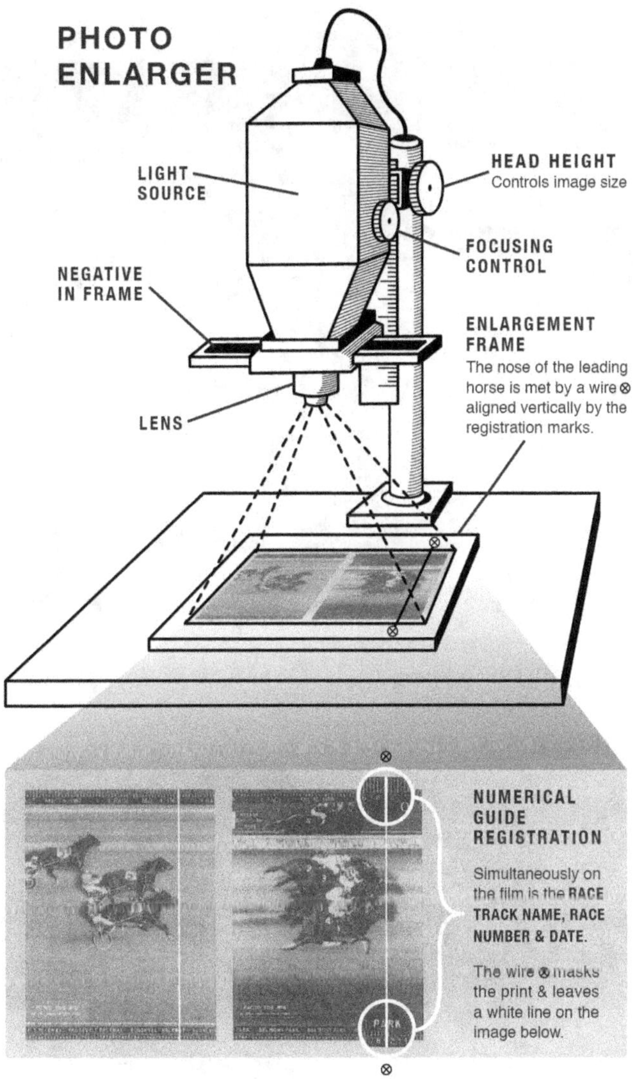

Figure 4.11 Tyler Barnes, *From Darkroom to Photo Finish*, 2018. © T. Barnes Graphics.

crews frequently appear out of synchronicity in photo-finish images (Figs. 4.14 and 4.15). For instance, in the 1968 Olympic finals, teams were rowing in perfect synchronicity, and had rigid, functional oars. But because the crew and oars were constantly moving in different directions as the boat passed the camera exposure space at the finish line over a duration of time, crew members' oars appear to be curved, and the crew itself appears to be out of synchronicity in the resulting photo finish.[73]

This rupture can make some viewers very uncomfortable. For example, in 2004 year the *Toronto Star*'s editors felt compelled to provide their readers with a lengthy

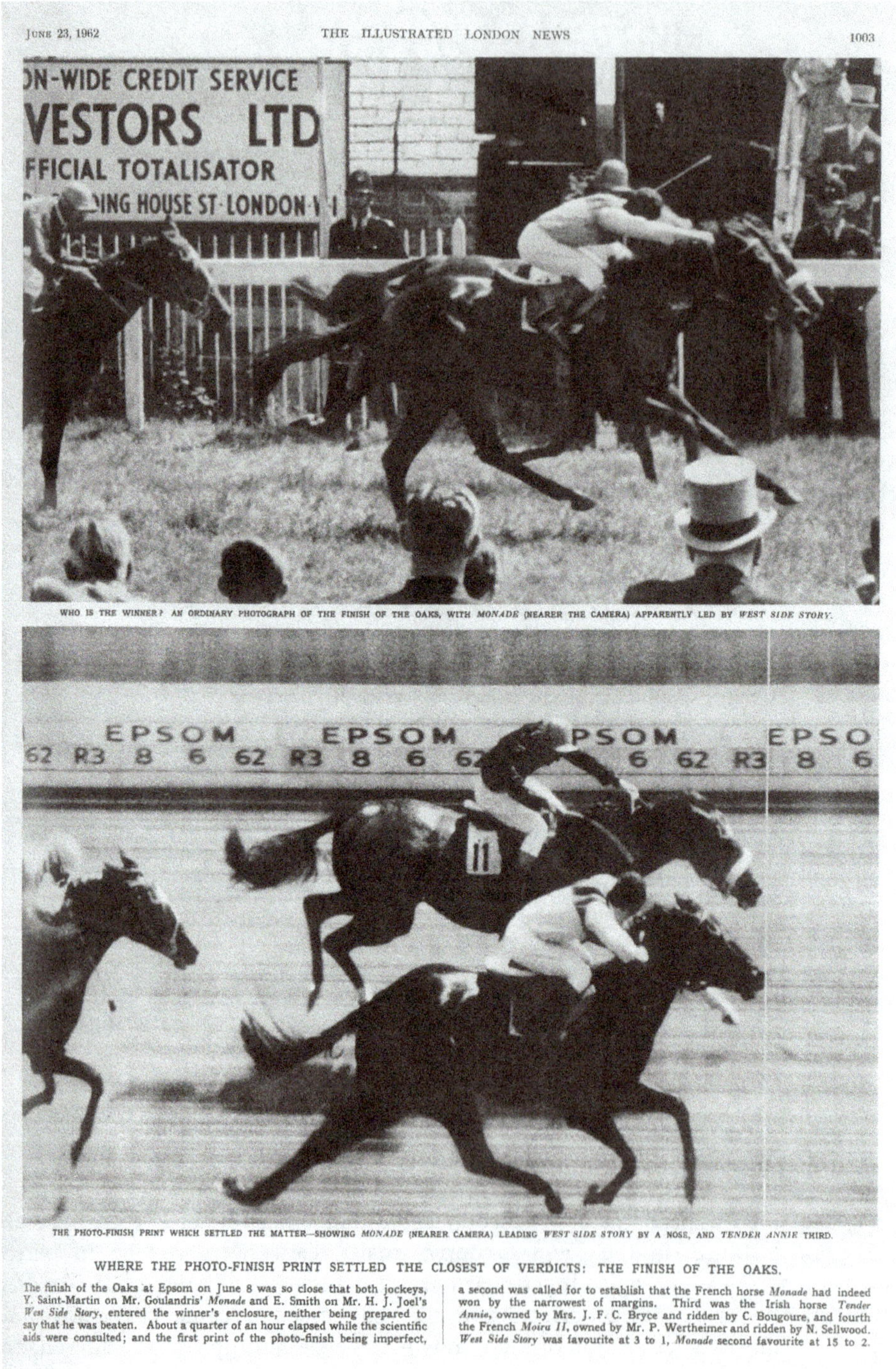
June 23, 1962 THE ILLUSTRATED LONDON NEWS 1003

WHO IS THE WINNER? AN ORDINARY PHOTOGRAPH OF THE FINISH OF THE OAKS, WITH *MONADE* (NEARER THE CAMERA) APPARENTLY LED BY *WEST SIDE STORY*.

THE PHOTO-FINISH PRINT WHICH SETTLED THE MATTER—SHOWING *MONADE* (NEARER CAMERA) LEADING *WEST SIDE STORY* BY A NOSE, AND *TENDER ANNIE* THIRD.

WHERE THE PHOTO-FINISH PRINT SETTLED THE CLOSEST OF VERDICTS: THE FINISH OF THE OAKS.

The finish of the Oaks at Epsom on June 8 was so close that both jockeys, Y. Saint-Martin on Mr. Goulandris' *Monade* and E. Smith on Mr. H. J. Joel's *West Side Story*, entered the winner's enclosure, neither being prepared to say that he was beaten. About a quarter of an hour elapsed while the scientific aids were consulted; and the first print of the photo-finish being imperfect, a second was called for to establish that the French horse *Monade* had indeed won by the narrowest of margins. Third was the Irish horse *Tender Annie*, owned by Mrs. J. F. C. Bryce and ridden by C. Bougoure, and fourth the French *Moira II*, owned by Mr. P. Wertheimer and ridden by N. Sellwood. *West Side Story* was favourite at 3 to 1, *Monade* second favourite at 15 to 2.

Figure 4.12 Unknown Photographers, *Comparing a Single-Exposure Race Finish Photograph to a Photo Finish, The Illustrated London News*, June 23, 1962, pg. 1003. © The Illustrated London News Ltd./Mary Evans.

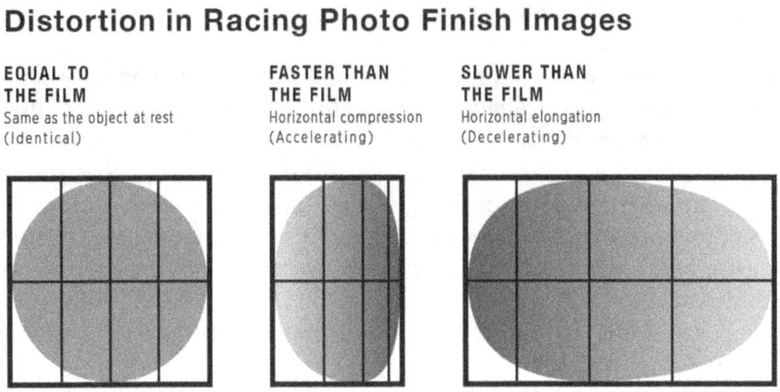

Figure 4.13 Tyler Barnes, *Diagram of Distortion in Photo Finish Images*, 2018. Based on diagram by Andrew Davidhazy. © T. Barnes Graphics.

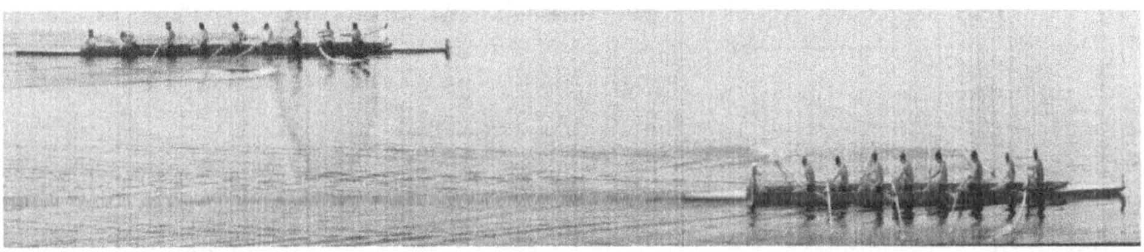

Figure 4.14 Detail, *Race Finish, Eight Oars and Coxswain*, Photo Finish, Mexico City, Mexico, 1968. West Germany took first place. The third- and sixth-place winners (the Soviet Union and the United States, respectively) appear here. In a racing photo finish, which records race finishes sequentially, rowers who were in synchronicity appear out of tempo with each other, and oars that were moved in and out of the recording field appear to be "bendy." © International Olympic Committee.

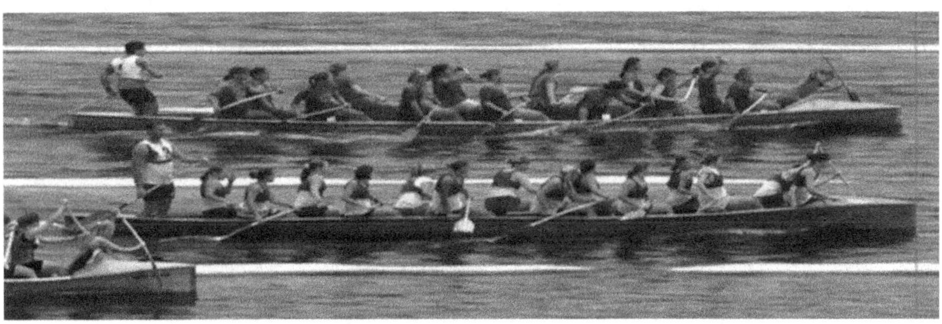

Figure 4.15 Detail, *Women's War Canoe Race*, n.d. © Lynx System Developers, www.finishlynx.com/allsports/rowing/en_US/rowing_images.html.

apology for running what they later felt was a deceptive photo-finish rowing match image from the Olympic Games:

> it wasn't a photo as most of us know photos. Just an image of four British rowers nosing out four Canadians at the finish line – a picture of time created through the miracle of Scan'O'Vision ... But if you look closely, the image isn't real ... Oars no longer are synchronized. And the winning boat seems longer. That's because the composite image only shows you how various boat and rower parts appeared as each crossed the finish line. Goodness, this is complicated. But here's the point. *The Star* published an artificial Scan'O'Vision image but told readers it was a photo. That's a no-no. *The Star's* policy manual ... warns editors not to use manipulated images, composites, graphics and montages.[74]

As the *Toronto Star*'s editors suggest, photo finishes require viewers to imagine a new relationship between photography and time, and between truth and abstraction. But instead of providing readers with an explanation of how the photograph was made and arguing for its faithful indexicality to the truth of the order of a race finish, editors condemned the image and apologized to their audience.

Yet more distinct examples of photo-finish distortions are apparent in photo finishes from track and field events. Because runners commonly step on the finish line, or may advance their legs at the speed of the film's movement, photo-finish images often feature missing, stretched, and deformed human legs (Fig. 4.16). The likenesses we see in a photo finish do not match our knowledge of the appearance of the represented subjects.

But photo-finish images, for all of their abstractions, *do* represent the temporal intervals between objects with absolute fidelity (Fig. 4.17). No matter how elongated or compressed an object might appear, the relative intervals between it and other objects will be recorded with precision within 1/2,000th of a second.[75] An object cannot appear on film until the precise moment when it moves past the

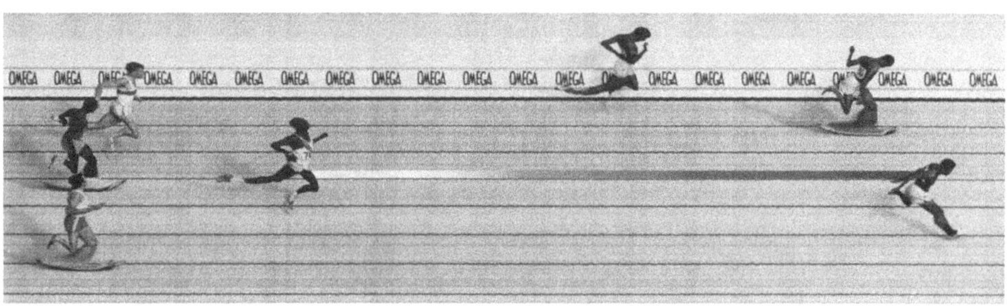

Figure 4.16 Race Finish, Lee Evans Wins the 400-Meter Men's Final, Olympic Games, Mexico City, 1968. Second-, sixth- and seventh-place runners (Larry James of the USA, Tegegne Bezabeh of Ethiopia, and Andzej Badeński of Poland, respectively) each planted a foot on the ground when they reached the finish line. This caused their feet to be present in the recording field for a longer duration of time, and accounts for their ski-shaped-feet in the photo finish. © International Olympic Committee.

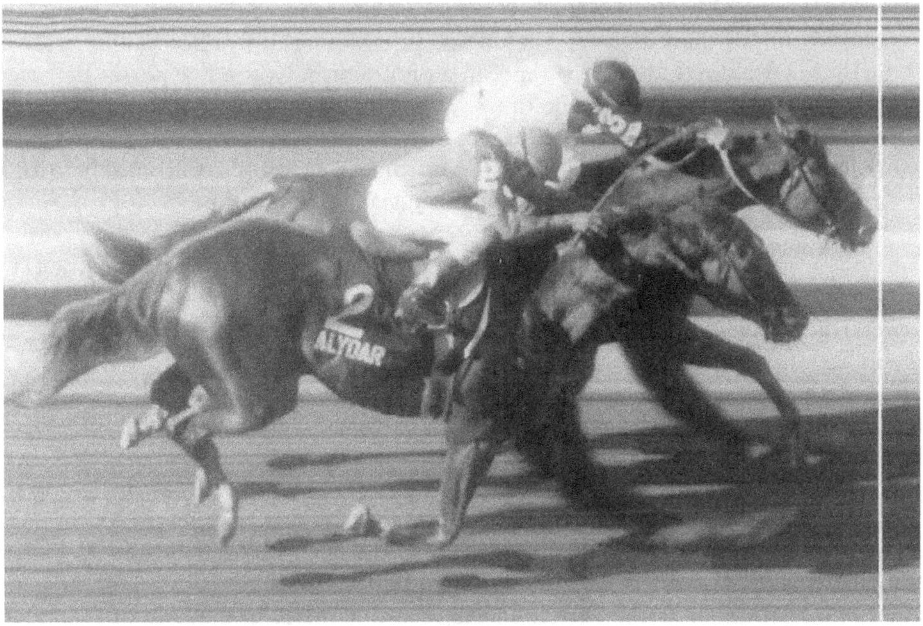

Figure 4.17 D&S Photo, detail, *Photo Finish, Alydar vs Affirmed, June 10, 1978*, copy print, 8 in. x 10 in. (20.3 cm x 25.4 cm). Courtesy of Don Morehouse, D&S Photo.

finish line. *No amount of distortion can make an object reach the aperture's recording field any sooner.* This is why a photo finish is upheld as absolutely truthful to the order and presence of racers at a finish line.[76] Thus, the photo finish reconciles what Barthes called photography's peculiar temporality to its capacity for a certain kind of realism: "false on the level of perception, true on the level of time."[77] Specifically, a photo finish is regarded as fail-safe, and as absolutely indexical to *one* kind of time (determining the order of finishes in a finite, linear duration) – but it does not provide an accurate account of the appearance of the racers. As such, a photo finish displays another phenomenon described by Barthes: that a photograph's power to certify presence may exceed its power to represent appearances accurately.[78] Nonetheless, a photo finish is an abstraction that enjoys the truth value of indexicality. It is consistent with Charles Sanders Peirce's own notion of the "index," which leaves the photograph room for contingency, so that it can slip in and out of semiosis.[79]

The photo finish also prompts us to question photography's relationship to veristic "truth," as Joel Snyder and Neil Walsh Allen have suggested.[80] In the article "Photography, Vision, and Representation," Snyder and Allen cite the photo finish as an example of one of several photographic practices that complicate essentialist claims that the medium has a fixed relationship to time and to "the real."[81] They note that the strip-photography photo finish has a peculiar manner of expressing time and space that departs from the medium's "instantaneity"-conditioned conventions:[82]

> Of course, once we know how a photofinish picture is made, it upsets *us*. We are accustomed, when we see five horses occupying five different positions in a photograph to think that we are looking at a picture of five horses that were all in different places at the same time. In a photofinish, we see five horses that were at the same place at different times. When we look at the nose and tail of a single horse in the picture, we are still looking at things which were recorded as they occupied the same place at different times. As we move from left to right across the picture, we are not looking at distance, but at time.[83]

Time in a photo finish is simultaneously absolutely indexical and abstract (or "symbolic" in Peirce's terms), and it suggests that photography as a whole cannot universally be defined accurately by any fixed relationship to time, but by the possibility of *many* – collapsed into a single image.

To further complicate matters, when racing subjects are moving in different directions as they cross the finish line (Fig. 4.18), they are recorded as moving in the same direction – the direction in which the film or recording field has been calibrated to move. In this photo finish, the runner was moving from the left to the right, in accord with the direction of the recording field, while the cyclist was moving from the right of the recording field to the left of it. This produces blatant distortions from known reality, although the photo finish does record, with absolute accuracy, each runner as they arrive in the recording field. In the case of this photo finish of the runner and bicyclist who are moving toward the finish-line field from different directions, the runner appeared within the "finish-line" recording area first, and was recorded as the "winner" – or, the first racer to appear on the right of the resulting image. (Any photo finish can be horizontally "flipped" to reverse this effect, although in the racing industry it is standard to show racers moving from left to right, with the first-place finisher(s) being the racer(s) farthest to the right.) Perplexingly, the shadows remain unchanged, and are the only hint that the racers were moving in different directions during the making of this image.

Such distortions have made strip photography an enticing medium in which to experiment with the fluid, and even sculptural, visual representation of time itself. In

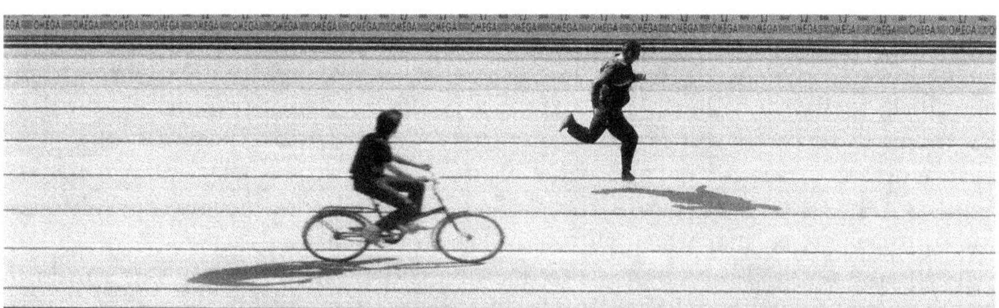

Figure 4.18 Omega Corporation/International Olympic Committee, Further Complications Posed by Strip Photography, n.d.: If moving objects are photographed by a strip camera while they are moving in opposite directions, the photo-finish camera will record them as if they were moving in the same direction. The direction of their shadows will still be correct, however. © International Olympic Committee.

Andrew Davidhazy's photograph of *Carlissa* (2009), which appears on the cover of this book, a wealth of distortions from the appearance of known reality appear when a strip-photography-related technology is used to sequentially photograph a revolving portrait.

Strip Photography in Its Technological and Historical Context: The Panorama

A strip-photography photo finish departs from "conventional," fraction-of-a-second, full-aspect-ratio photographs in three ways. It is (1) a time-lapse exposure that is (2) made through a limited, altered aperture field, (3) using a mobile, light-sensitive recording surface.[84] As the combined product all three conditions, the photo finish is a descendant of peripheral ("rollout") photography and of swing-lens/rotating-body panoramic practices.[85]

The first of these practices to emerge was panoramic photography, which features 100- to 360-degree-wide views of a subject that are recorded on long plates or strips of film using mobile lenses or plates/film.[86] Panoramic cameras fit into three broad categories: (1) wide-angle-lens cameras that feature instantaneous exposures (such as a 35-millimeter camera equipped with a fish-eye lens); (2) swing-lens cameras such as the Megaskop used by Friedrich von Martens (Fig. 4.19), in which the camera's body does not move, but the lens swings in an arc of up to 270 degrees, exposing the film gradually; and (3) cameras with rotating bodies and stationary lenses, such as the Kodak Cirkut Camera, which allow the entire camera to pivot and cover a span of up to 360 degrees.[87] Within these three types are many variations, including panoramas comprised of multiple exposures montaged-together, as Joachim Bonnemaison has suggested.[88] Swing-lens and rotating-body cameras both have scanning slits that allow for the sequential exposure of the film or plate to light over an elapsed duration of time.[89]

These varied "scanning camera" panoramic techniques evolved from photographers' efforts to extend the limited aspect ratios of conventional photographic views. Perhaps the most popular of such images was made by Antoine Claudet, who photographed views of London from atop the Duke of York's column in 1842. He joined several separate daguerreotype plates to create a panoramic image that replicated the extended views of panoramic paintings. C. F. Sargent took Claudet's fused daguerreotype images and made them into a continuous view using sixty connected boxwood blocks, and engraver Ebenezer Landell prepared the image published in the January 7, 1843, issue of the *Illustrated London News*. The panorama was a "gift" for customers who had purchased six-month subscriptions to the new weekly publication, and it contributed to the immediate success of the newspaper. Subscriptions rose from 26,000 to 66,000 from May to December of 1842.[90]

According to Stephan Oettermann, the pre-photographic panorama and its many variants – cosmoramas, dioramas, neoramas, cycloramas, and moving panoramas – "came to epitomize an international hunger for physically, geographically, and historically extended vision" in the early to-mid-nineteenth century.[91] To Oettermann, the desire for an "extended vision" to the displacement of formerly fixed notions of time and space that was encouraged by the sciences and by various developments of the modern age, including airborne travel, rapid transportation, and instantaneous communication.[92] He warns, however, that this new popular manner of seeing the world was foreign to human biology, but wholly "modern" in its genesis:

> It [the panorama] is the vision demanded by the natural sciences, which were developing rapidly at that time. The panorama is both a part of this process of transforming human vision and a product of it; as an art form, it is a response to a way of seeing that its viewers are already presumed to have, while at the same time it instructs them how to see in this manner. The discovery of the horizon reflects the historical experience that the known world is contained within it and an unknown world begins beyond it; both worlds should be explored ... Even the feeling of giddiness that accompanied this discovery is simulated in the panorama. According to contemporary accounts, a great many "ladies with delicate nerves" and "young dandies" became seasick at the first exhibitions.[93]

Maarten Vanvolsem has suggested that the panorama itself mimics the body's movement through the landscape; only it also adds an unfixity of vanishing points that disorients the body and sensorium, accounting for the vertigo effect noted by Oettermann.[94] Likewise, Sigmund Freud also provided a biological explanation for the operations of sensory overstimulation, and the degree to which they have become conditioned in the observer:

> as a result of the ceaseless impact of external stimuli on the surface of the vesicle, its substance to a certain depth may have become permanently modified, so that excitatory processes run a different course in it from what they run in the deeper layers. A crust would thus be formed which would at last have been so thoroughly "baked through"; by stimulation that it would present the most favourable possible conditions for the reception of stimuli and become incapable of any further modification.[95]

John Girdner also identified what he saw as a new social disease afflicting all urban dwellers, "Newyorkitis." This uniquely modern psychological condition caused a "rapidity and nervousness and a lack of deliberation in all movements," and was caused by modern stimuli's high-speed, multi-layered assault on a previously linear, slower-paced perception of the pre-modern age.[96] In a medical study, Willy Hellpach specifically tied this new state of "nervousness" to the increasing rapidity of communication and transportation in the nineteenth century which caused an "overwhelming increase of normal mental processes."[97]

In *Technics and Time*, philosopher Bernard Stiegler suggests that technological advance produces "a divorce ... between the rhythms of cultural evolution and the rhythms of technical evolution."[98] This produces a situation in which human biology rebels and "[t]echnics evolves *more quickly* than culture."[99] Stiegler further argues that this tension results in a condition in which "there is both advance and delay, a tension characteristic of the extending ... that makes up any process of temporalization."[100] For instance, the decades surrounding the invention of photography brought a flurry of innovation, followed by a brief lull.[101] The embodiment and interpretation of time perceptions by the observer, Jonathan Crary argues, renders temporality unfixed, subjective, and contingent in an age of modernity.

These tendencies, he argues, are defining characteristics of the modern age – and they define an unprecedented "modern" consciousness.

Despite the jarring effects of such technologies upon some viewers, enthusiasm for an extended vision of the landscape – exemplified by the popularity of Claudet's panorama – inspired several innovators to develop single-exposure cameras that might make the fusion of plates unnecessary. Joseph Puchberger received a patent for a swing-lens panoramic camera that used curved daguerreotype plates in 1843. His camera employed curved daguerreotype plates that were 19 to 24 inches long, and had an 8-inch focal-length lens. Puchberger's photographs spanned a 150-degree arc. The same year, Calvert Richard Jones developed a "double picture" panoramic camera, which he pivoted to make two separate, successive, overlapping calotype exposures of a single subject.

Von Martens patented his Megaskop camera – which made 150-degree, swing-lens photographs on 4.7-inch-by-15-inch curved daguerreotype plates that later were flattened for viewing – in 1844.[102] The panning speed of the Megaskop's lens was controlled by gears to achieve less spatial distortion and to assure von Martens of an achieving an evenly exposed plate. His resulting panoramic images offer a unique collision of naturalism and abstraction. For instance, one such image, von Martens's *View of Paris* (1845) (Fig. 4.19), provides a laterally reversed, black-and-white image of the city scene it represents. Although von Martens's camera appears in the image to have been positioned near today's Musée d'Orsay (7th arrondissement), it actually was placed on the Left Bank near the Gare d'Austerlitz (13th arrondissement), just south of the foregrounded Austerlitz Bridge. In von Martens's panoramic view, Parisians saw a scene that did *not* resemble the city they knew. Yet the daguerreotype indeed represented an image that was recognizably *Paris* – from the decorative details on the side of the Austerlitz Bridge to the distant proud rise of Notre Dame on the Ile de la Cité. Moreover, the reflective metal daguerreotype plate enabled viewers to see a large mirror image of their own face surrealistically superimposed in comparably large scale, in the laterally reversed, but highly detailed, black-and-white Paris cityscape. As Paris rests in eternal stasis in photographic grayscale, the viewer's own reflection discordantly enjoys motion and full-color realism.

Further tensions between realism and abstraction also occur when arc-shaped photographic plates are used in swing-lens cameras such as von Martens's. The rounded photographic plate coincided with the shape of the swinging lens's circular trajectory. But after arc-shaped plates are flattened for display, many features of the image are represented as more arc-shaped than they are in actual appearance. For instance, although the Right Bank of the Seine in von Martens's *View of Paris* is fairly straight, it curves outward dramatically at the left edge of the image. The Austerlitz Bridge sags slightly.

Like its later descendant, the photo finish, the panorama also represents a finite duration of light captured sequentially, in linear time. As panoramic photography was adapted for taking group portraits in the late nineteenth to early twentieth centuries, photographers anticipated these distortions, and compensated for them accordingly. In order to produce a final portrait in which the subjects appeared to be standing in straight horizontal rows, photographers using revolving cameras (such as the Kodak Cirkut Camera), and arranged the group in a large circle, with the camera positioned at its center. Often, they drew circular chalk marks on the ground and built towers at the site of the portrait to aid in the image's execution (Figs. 4.20 and 4.21).[103] The

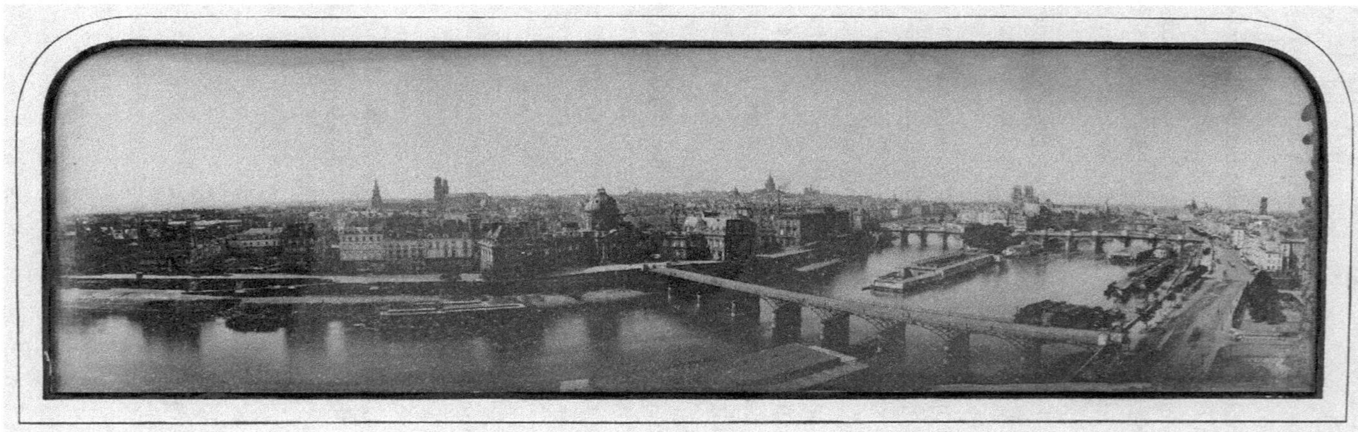

Figure 4.19 Friedrich von Martens, View of Paris, 1845. Courtesy of the George Eastman Museum.

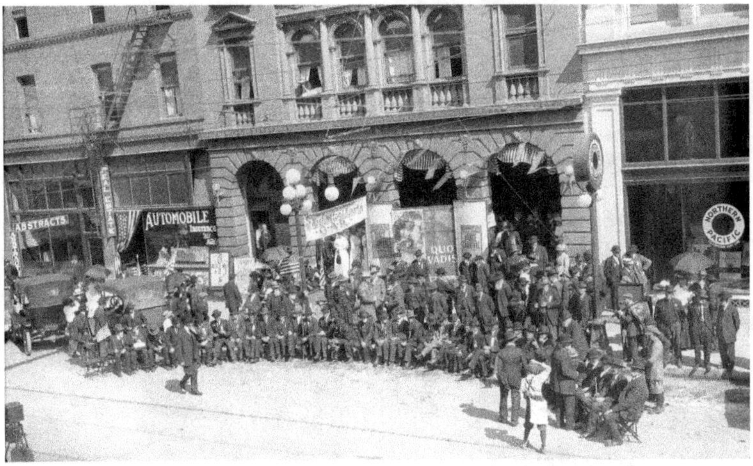

Figure 4.20 Sarah Quackenbush, *Setting Up Panoramic Photograph, Civil War Veterans, Bellingham, Washington*, 1914. © Whatcom Museum of History and Art, Catalog # 1996.10.4972.

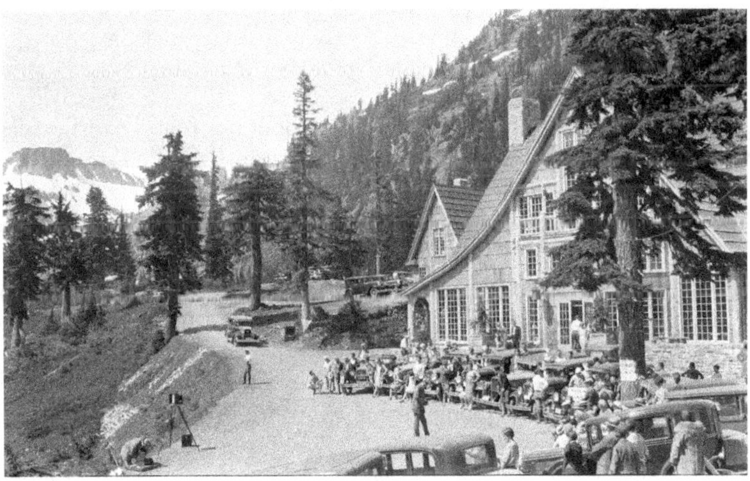

Figure 4.21 Bert W. Huntoon, *Posing for Cirkut Camera, Gilmore Oil Group at Mt. Baker Lodge, Bellingham, Washington*, c. 1930. © Whatcom Museum of History and Art, Catalog # 1996.10.2794.

Cirkut was mounted on a tripod, and the entire camera rotated in one direction while the film advanced at the same speed in the opposite direction. Cameras came in five sizes (#5, 6, 8, 10, and 16), which corresponded to the size of the film (which ranged from 5 inches by 3 inches [#5] to 1 foot 4 inches by 20 *feet* [#16]).

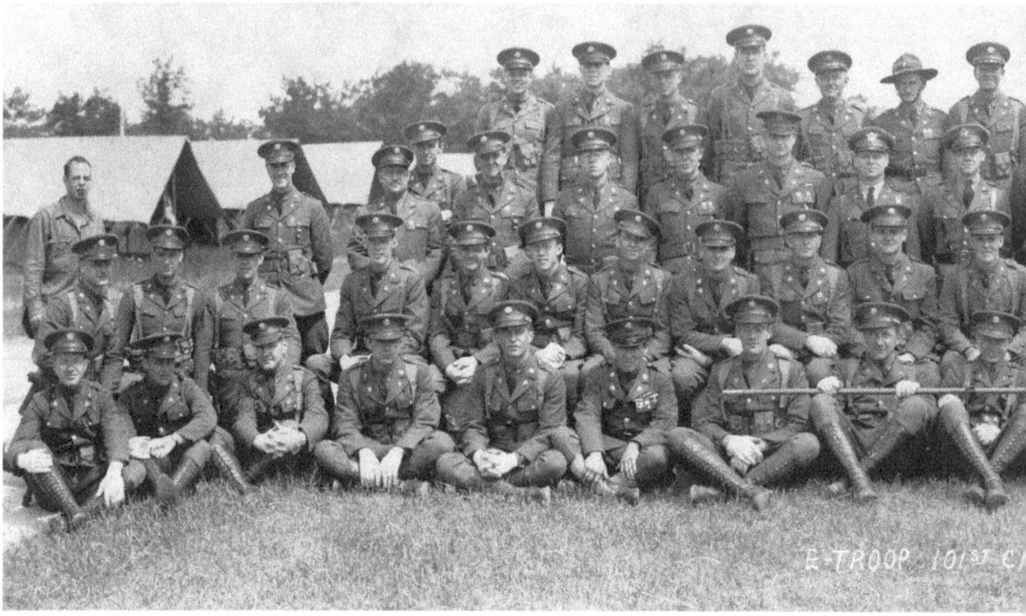

Figure 4.22 Robert B. MacKay, *E-Troop, 101st Calvary, Pine Camp, NY, Double-Ender Cirkut Camera Image*, 1937.

The resulting panoramic photograph, however, does not reveal overt evidence of the group's positioning in a semi-circle. Instead, subjects appear to be standing in a somewhat straight line. The image is truthful to the *presence* of its subjects rather than to the *appearance* of the group's arrangement. Rounded space is flattened and reconstituted by photography, as space also was distorted by other inventions of the modern age. The "unreal realism" of panoramas required "a considerable suspension of disbelief," Miles Orvell argues.[104] Yet "the viewer accepted the … [panorama] as a kind of model, an artificial construction made out of 'real' pictures."[105]

The subjects of panoramic photographs occasionally took playful advantage of the camera's extended-duration, sequential-time exposure. In the panoramic photograph Robert B. MacKay, *E-Troop, 101st Calvary, Pine Camp, NY, Double-Ender Cirkut Camera Image* (Fig. 4.22), a subject positioned near the end of the circle, Walter Devereux, ran behind the camera to uncannily appear on two ends of the resulting group portrait.[106] This "double-ender" panoramic portrait questions conventional expectations fostered by instantaneous photography: that a photograph represents spatial relationships in a fraction of a moment of time. Instead, subjects such as Devereux have the opportunity for more than *one* singular presence and relationship to space in the photograph's represented time. Devereux here realizes a distinctively "modern" desire: the sensation of the annihilation of the perceived rules of space and time. He playfully defies the laws of Euclidean geometry and rationalist seventeenth- and eighteenth-century Newtonian physics, which dictated that a subject was afforded *one* physical presence in space in an instant of linear time.

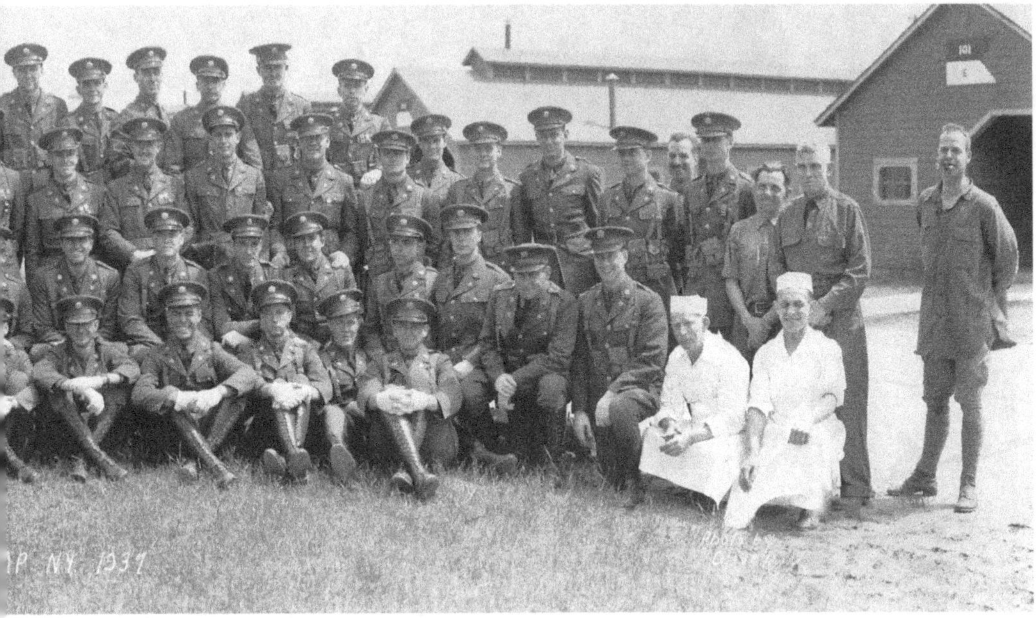

Device-altered photography reflects an age in which understandings of time and space radically and creatively have been reconfigured.

Oettermann attributes the panorama's extension of the metaphorical and literal horizons of the photographic aspect ratio to modernity's expanded conceptions of the temporal and spatial matrix. Swift new forms of transportation such as the railroad, automobile, and airplane, he argues, "reduced the sensation of being closed in by the horizon," so that "[t]o the modern eye the horizon no longer appears as a painful barrier. It has become little more than a metaphor, 'pierced' in a thousand places by supersonic airplanes and satellite television."[107] The panorama, as Oettermann and Merritt Ierley have argued, was hailed as an ideal medium for reflecting the public's perceptions of temporal and spatial fluidity in an age of new order, industrial and technological modernity.

Peripheral Photography as Photo-Finish Precedent

Photo-finish technology also is related to peripheral, or "rollout," photography (Fig. 4.23). In peripheral photography, the camera may or may not move as it continuously records the 360-degree surface of the outer periphery of a cylindrical or near-cylindrical subject on a flat, light-sensitive photographic surface.[108] To make a peripheral image, a subject is first placed on a turntable. Film inside of a camera is calibrated to move at the same speed as the subject. As the film slides through the camera and the subject revolves, an image is recorded

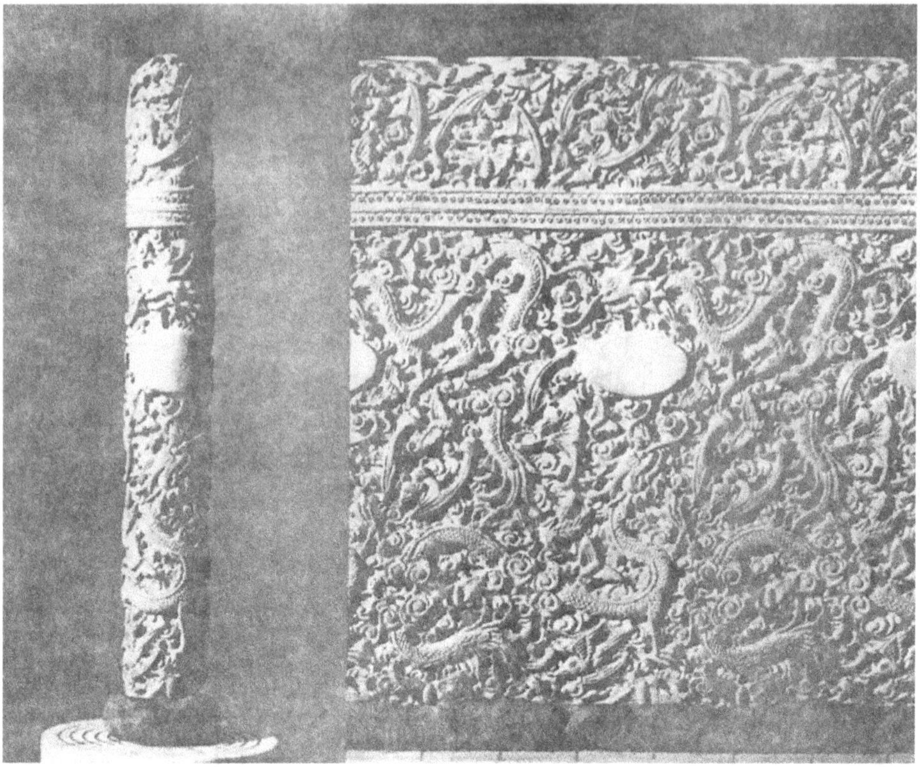

Figure 4.23 Research Engineers, Ltd., *Rollout Photography/Peripheral Photography*, Company literature. 1950. Public Domain.

progressively as light falls upon the film through a thin vertical, slit-shaped aperture. The end result is a long, thin image of the subject's entire surface (such as the surface of the tire, Fig. 4.24). What was once a three-dimensional cylindrical surface is translated into a long, flat image. Like the strip-photography photo finish, swing-lens panoramas, rotating-camera panoramas, and both categories of peripheral images all feature the gradual recording of a subject over elapsed, linear sequential time.

Although little is known about the origins of the peripheral camera in the nineteenth century, a report published in May of 1895 states that Arthur Hamilton-Smith from the Greek and Roman Antiquities department at the British Museum designed an instrument for photographing the painted surfaces of the ancient pottery works from its collection.[109] The camera system later was expanded for use by other departments at the British Museum.[110] Before the peripheral camera was developed and introduced to the museum by Hamilton-Smith, art historians, archaeologists, and anthropologists regularly made relief "rubbings" of the surface contours of sculptures, pottery, and architecture-embedded carvings, in order to view all of their surface ornamentation at once. This enables the study of the object's iconography, stylistic affinities, and pictorial tropes including contour rivalry.

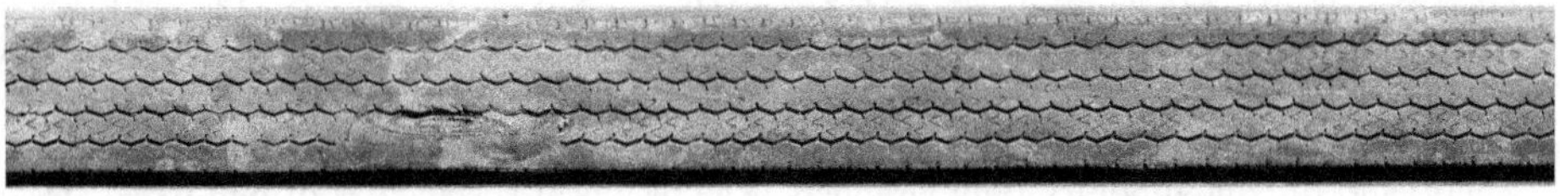

Figure 4.24 Andrew Davidhazy, *Worn Spot on the Surface of an Automobile Tire*, Peripheral Photograph. 1975. © Andrew Davidhazy.

To fully appreciate the peripheral "rollout" image's translation of reality to an image, viewers are prompted to visualize a three-dimensional object such as a vase or a relief carving on a column, and imagine that its surface ornamentation is translated into a flat, often scaled-down, two-dimensional image. This process of abstracting three-dimensional objects into two-dimensional form is not unique to the peripheral photography. Swing-lens and rotating-body panoramic images, as mentioned previously, also distort the appearance of the spatial arrangement of their subjects while also offering unique abstractions resulting from the images' linear, extended-duration exposures.

Conclusions

Highly experimental and technical in nature, device-altered photography exemplifies and enables the medium's insistent struggle to master time's passage, to extend the capabilities of unaided human vision, and to find photography's time-depiction loopholes and conquer them. This set of photographic practices is scientifically positivist in spirit, but offers a paradoxical, complex blend of realism and abstraction. Perhaps fittingly, this set of photographic practices emerged as a flurry of inventions reinforced this sensation of temporal/spatial fluidity. Strip photography and its adjacent practices are a uniquely appropriate means for expressing states of contradiction and flux, of reality and unreality, of absence and presence, and of indexicality and abstraction – opposing concepts which form and inform consciousness. Photography contributes a distinct medium for expressing of what it means to live in a time of industrial and technological modernity – with all of its exhilaration, scientific and technological advance, and the resulting anxieties it stirs over the forced and synthetic nature of the acceleration of embodied temporality. Photography specifically provides a means for manipulating the recording of time's irregular passage, and expressing not how time *appears*, but an expression of how it *feels*.

Notes

1 Some parts of this chapter first appeared as a shorter essay in: Kris Belden-Adams, "A Wrinkle in Time," *Cabinet*, Vol. 25 (Spring 2007): 24–31.
2 Don Morehouse, Tour of Aqueduct Race Track Photo Finish Booth, Interview by Kris Belden-Adams (March 11, 2005).
3 When the photo finish is discussed at all, it finds its audience in horse-racing magazines, users' manuals, and in technical journal articles by practicing artists/scholars such as Andrew Davidhazy, Maarten Vanvolsem, and Tania Ruiz Gutierrez: Tania Ruiz (Gutierrez), "Études sur le Temps et l'Espace dans l'Image en Mouvement: Tissage Vidéo, Objets Spatio-Temporels, Images Prédictives et Cinéma Infini" (PhD dissertation, Université Paris

1, 2004); Maarten Vanvolsem, *The Art of Strip Photography: Making Still Images With a Moving Camera* (Leuven: Leuven University Press, 2011). Andrew Davidhazy has produced the bulk of scholarship on the various approaches to device-altered photography, and he actively experiments with these techniques in his studio practice. In effect, Davidhazy's contributions have introduced this set of practices to the academic community, and to contemporary photographers. The technical specificities of film-based streak, strip, panoramic, peripheral, stroboscopic, and synchroballistic photography are discussed in several articles by Andrew Davidhazy of the Rochester Institute of Technology, most of which are also available on this website: http://people.rit.edu/andpph/articles.html. Several articles of particular relevance to this study include: Andrew Davidhazy, "Simplified Streak/Strip and Scanning Photographic Systems" (Orlando, FL: International Panoramic Photographers Association Conference, February 5, 1985), n.p.; Andrew Davidhazy, "A New Spin on Panoramic Photography," *Photo Educator International* (1992); Andrew Davidhazy, "Camera for Conical Peripheral and Panoramic Photography," *Current Developments in Optical Engineering and Commercial Optics*, Vol. 1168 (1989), available online: http://portal.acm.org/citation.cfm?id=1281505; Andrew Davidhazy, "A Simplified Stroboscopic System for Motion Pattern Photography," *Tech Directions*, Vol. 59, No. 7 (February 2000): 12–16; Andrew Davidhazy, "Synchroballistic Photography," "Alternative Velocity Measurement Techniques in Synchroballistic Photography" [Online essays], http://people.rit.edu/andpph/articles.html (accessed October 15, 2009), n.p.

Omega Sports Timing (now owned by Swatch) recounts its corporation's history of time keeping at the Olympic Games in: Jean-Pierre Bovay, *Le Sens du Temps: Time Photographs and Perspective* (Bienne, Switzerland: Omega Sports Timing Corporation, 1988). Bovay's book includes a brief, non-scholarly account of the development of photo-finish technology, in the service of providing an historical context for the Omega Sports Timing company's contributions.

4 Andrew Davidhazy, "Interview by Author" [Email] (2005).

5 The "fixed, rectangular space" of a photograph made with a 35-millimeter camera without alterations to its focal plane has a 3:2 width-to-height aspect ratio. This ratio has become a standard for photography – both film-based and digital: "Image Aspect Ratios: Common Aspect Ratios for Digital Images, Photography, Cinema, and Television," http://home.comcast.net/~igpl/Aspect.html (accessed August 23, 2007).

6 These categories are by no means an exhaustive representation of all photographic practices which depict device-altered time. Some other seemingly infinite variations and their applications are described in technical detail in: Ron Jegerings, "Photography: Motion Effects Techniques," *Darkroom & Creative Camera Techniques*, Vol. 14, No. 1 (1993): 20.

7 Muybridge applied for a United States Patent for a "Method and Apparatus for Photographing Objects in Motion" in June 1878, and patent no. 212,865 was granted to him in March of 1879: Eadweard Muybridge, "Improvement in the Method and Apparatus for Photographing Objects in Motion," United States Patent Office (United States of America: 1879). Muybridge first began photographing a horse in motion for Leland Stanford in the summer of 1872, and published his images in 1878. For the most part, they were met by worldwide acclaim. By all written accounts, horse-racing enthusiasts especially praised Muybridge's studies as "unerring": *The Spirit of the Times* (June 3, 1882): 513.

8 "Progressive Photography," *The Spirit of the Times* (October 11, 1879): 252. The author of this passage refers to Muybridge's experiments as "electro-photography," because Muybridge's shutters were operated by electricity.

9 Eadweard Muybridge, "A Dead Heat," *Nature*, Vol. 26 (1882): 81.

10 André Bazin, "The Ontology of the Photographic Image," *Film Quarterly* Vol. 14, No. 4 (1960): 4–9.

11 André Bazin, *What Is Cinema?*, Vol. 1 (Berkeley, Los Angeles and London: University of California Press, 1967), 13.

12 Gerald Stanley Lee, *The Voice of the Machines* (New York: Mount Tom Press, 1906), 58.

13 Lee, *The Voice of the Machines*, 58.

14 Scott McQuire, *Visions of Modernity: Representation, Memory, Time and Space in the Age of the Camera* (London: Sage, 1998), 34–35. The author's spelling has not been changed.
15 Howard Bessner, "Elements of Modern Consciousness." www.gseis.ucla.edu/~howard/Papers/consciousness.html#component (accessed March 16, 2008), n.p.
16 "Is It a Push or a Pull?" *The Blood-Horse* (March 22, 1947): 833.
17 Jerome P. O'Neill, "The History of Race Photo Finish Photography: Interim Report of the Photo Finish Research Project" (Rochester: 1972): 2–3.
18 Using stereoscopic lenses allowed Muybridge *two* images to choose from for each frame of *The Horse in Motion*. This decreased Muybridge's chances for technical error, although it also implies another level of editorial decision-making in Muybridge's selection of the "best" images for the series. Additional temporalities are implied not just by the instants of the photographs' taking and the time represented by each image – and the series as a whole. The time of Muybridge's technical experimentation, the time needed to process the images, the time he used to select images, and the time needed to compose the series also are implied in the final image. Yet more temporalities are implied at the time of viewers' reception of the images. The material aging of Muybridge's images also has temporal implications. Muybridge's interventions therefore complicate Bazin's earlier statement that photography is an impassive mechanical process that is therefore obligated to provide a truthful visual record of the temporal and spatial existence of objects within a fraction of a second of time.
19 Muybridge published his motion studies in these volumes: Eadweard Muybridge, *Animals in Motion* (London: Chapman and Hall, 1899); Eadweard Muybridge, *Animal Locomotion: The Muybridge Work at the University of Pennsylvania* (Philadelphia: J. B. Lippincott Co., 1888).
20 Jerome P. O'Neill argues that Muybridge may have exaggerated his shutter speeds: "Muybridge claimed shutter speeds as high as 1/5,000 second with wet collodion plates, but sufficient exposure would have been a problem. In his later work he found that 1/600 to 1/800 second was adequate to 'stop' the stride of a horse": Jerome P. O'Neill, "The History of Race Photo Finish Photography," 10. Muybridge was not the only photographer experimenting with shutter speeds at this time. German photographer Ottomar Anschütz patented a focal-plane shutter in 1888 which could open and close as quickly as 1/1,000th of a second. For more on Anschütz's experiments, see: Keith James Laidler, *To Light Such a Candle: Chapters in the History of Science and Technology* (Oxford and New York: Oxford University Press, 1998), 108.
21 Placing the subject in front of a grid was Muybridge's way of lending the project scientific credibility, Marta Braun argues: Braun, "Time and Photography," in *Tempus Fugit: Time Flies*, ed. Jan Schall (Kansas City, MO: Nelson-Atkins Museum of Art, 2000), 134–135.
22 These gaps, perhaps, have a biological familiarity. They are similar to the breaks in the continuity of vision that we experience when we blink.
23 We also experience life this way. Although we cannot remember the event of our birth, we do remember our lives in progress, starting around three or four years of age. Presumably, we also consciously will experience the endpoint of our lives.
24 On March 6, 2008, I taught *Horse in Motion* to a group of animation majors at the Kansas City Art Institute. One of their assignments for an animation class was to re-animate Muybridge's *Horse in Motion* by filling in the gaps between these twelve photographs. My students unanimously agreed that every one of Muybridge's photographs – except for the last one – were sequential, plausibly separated by a half-second interval of time, and obeyed the mechanics of linear equine movement.
25 Marta Braun, "The Expanded Present: Photographing the Moment," in *Beauty of Another Order: Photography in Science*, Ann Thomas, ed. (New Haven: Yale University Press, 1997), 153. Braun also discusses several of Muybridge's altered motion studies in relationship to Marey's work in Braun, *Picturing Time*, 238–249.
26 Rebecca Solnit, *River of Shadows, Eadweard Muybridge and the Technological Wild West* (Middlesex, England: Viking Penguin, 2003), 221.

27 Josh Ellenbogen, "Photography and the Imperceptible: Bertillon, Galton, Marey" (PhD diss., University of Chicago, 2005); Josh Ellenbogen, "Authority, Objectivity, Evidence: Scientific Photography in Victorian Britain," *Studies in the History and Philosophy of Science*, Vol. 39 (Spring 2008); Josh Ellenbogen, "Camera and Mind," *Representations*, Vol. 101 (Winter 2008): 86–115.
28 Braun, "The Expanded Present," 153.
29 Braun, "Time and Photography," 133.
30 Braun, *Picturing Time*, 134–135. Braun distinguishes Muybridge's altered images of movement from Marey's scientific – and therefore more "truthful" – endeavors to record time's passage by representing movement.
31 Muybridge's scandalous life and his contributions to the media of photography and cinema have been the subject of many books and articles. Three fairly recent publications which look at Muybridge's exploration of instantaneous time include: Prodger, *Time Stands Still: Muybridge and the Instantaneous Photography Movement* (Oxford: Oxford University Press, 2003); Solnit, *River of Shadows*; Braun, "Time and Photography."
32 "Muybridge's Horses," *The Spirit of the Times* (June 3, 1882): 513.
33 Thierry De Duve, "Time Exposure and Snapshot," *October*, Vol. 5 (1978): 115.
34 Jonathan Crary, *Suspensions of Perception: Attention, Spectacle, and Modern Culture* (Cambridge, MA: MIT Press, 1999), 140. The year after he made *Horse in Motion*, Muybridge – who is acknowledged as one of the founding fathers of cinema – announced his design for an early version of a movie projector, which he called a "zoopraxiscope." This device presented replayable movement, another form of what Crary calls "reality effect." Crary's book *Techniques of the Observer* posed the possibility that photography did not, due to a technological connection to the camera obscura, embody perspectival systems of seeing from the Renaissance. Instead, Crary argues that medium might instead embody a condition of modernity that included a collapsing of the boundaries between observer and observed, subject and object, and representation and real: Jonathan Crary, *Techniques of the Observer: On Vision and Modernity in the Nineteenth Century* (Cambridge, MA: MIT Press, 1990).
35 Crary, *Suspensions of Perception*, 138.
36 Jules Portefin, "French Horse Painters of the Nineteenth Century," *British Racehorse* (November 1950): 395. Portefin argues that Théodore Géricault's passion for horses drove him to study art under Carle Vernet – who had a fiery disposition, but a great gift for painting horses. Portefin, "French Horse Painters of the Nineteenth Century," 394. Interestingly, French painter Théodore Géricault died from a horse-riding accident at the age of thirty-three.
37 Auguste Rodin, *L'Art: Entretiens Réunis par Paul Gsell* (Paris: Bernard Grasset, 1924), 86. Rodin's original quote is: "c'est l'artiste qui est véridique et c'est la photographie qui est menteuse; car dans la réalité le temps ne s'arrête pas." The above translation is mine.
38 Rodin's original quote is: "Ils critiquent Géricault parce que dans sa *Course d'Epsom*, qui est au Louvre, il a peint des chevaux qui galopent ventre à terre, selon l'expression familière, c'est-à-dire en jetant à la fois leurs jambes en arrière et en avant. Ils disent que la plaque sensible ne donne jamais une indication semblable. Et en effet dans la photographie instantanée, quand les jambes antérieures du cheval arrivent en avant, celles d'arrière, après avoir fourni par leur détente la propulsion à tout le corps, ont déjà eu le temps de revenir sous le ventre pour recommencer une foulée, de sorte que les quatre jambes se trouvent presque rassemblées en l'air, ce qui donne à l'animal l'apparence de sauter sur place et d'être immobilisé dans cette position. Or, je crois bien que c'est Géricault qui a raison contre la photographie: car ses chevaux paraissent courir: et cela vient de ce que le spectateur, en les regardant d'arrière en avant, voit d'abord les jambes postérieures accomplir l'effort d'où résulte l'élan général, puis le corps s'allonger, puis les jambes antérieures chercher au loin la terre. Cet ensemble est faux dans sa simultanéité; il est vrai quand les parties en sont observées successivement et c'est cette vérité seule qui nous importe, puisque c'est celle que nous voyons et qui nous frappe." This quote is published in: Rodin, *L'Art*, 86. The above translation is mine.

39 Irma B. Jaffe, "The Flying Gallop: East and West," *Art Bulletin*, Vol. 65, No. 2 (1983): 183.
40 Jaffe, "The Flying Gallop," 183, 185.
41 "The Instantaneous Process," *The Turf, Field and Farm* (June 18, 1886): 524.
42 O'Neill, "The History of Race Photo Finish Photography," 2–3.
43 Bazin, "The Ontology of the Photographic Image," 13.
44 Laurent Guido, *Between Still and Moving Images* (Bloomington: Indiana University Press, 2012), 10.
45 Guido, *Between Still and Moving Images*, 13.
46 "Editorial Notes," *Photographic Times and American Photographer*, Vol. 18 (1888): n.p. In some sources, Ernest Marx's last name is spelled "Marks." The name "Marx" appears more frequently in racing periodicals and in other primary source material. Consequently, this book uses that spelling of his name.
47 Eadweard Muybridge, *Scientific American*, Vol. 39 (1878): 239.
48 Most of the oldest photo-finish images have not survived because photographers and race tracks saw no reason to keep them. J. C. Hemment was an exception to this tendency. He assembled albums of his race-track images, which included photo finishes, photographs of jockeys and owners, race-track employees, and patrons. The Keeneland Library in Lexington, Kentucky, houses Hemment's albums.
49 *The Spirit of the Times* (June 26, 1886): 702.
50 Incidentally, in Hemment's own album of race-track photography, either Hemment himself or his assistant misidentified the year of this race (as 1891, instead of 1890). I am grateful to Kathy Schenck of the Keeneland Library for pointing this out.
51 Robert Harron, "Eye in the Sky," *Collier's Illustrated Weekly* (September 19, 1936).
52 Tracks invited members of the press to "tests" of the new devices, in order to convince the public of the reliability of the instantaneous photo finish. Journalists were not easily won over by improvements in the camera device: "Finish Photography," *The Blood-Horse* (June 20, 1936): 743.
53 "French Photography," *The Blood-Horse* (July 10, 1937): 53.
54 "Faraway," "Explanation of a Riot," *The Blood-Horse* (August 21, 1937): 241.
55 Ralph Powers, "Photography at the Race Track," *Photo Technique* (June 1939): 30.
56 "In the Wake of the News: Arlington Park," *Chicago Daily Tribune* (June 29, 1936): 19; "Roman Soldier to Campaign at Arlington Park," *Chicago Daily Tribune* (June 25, 1936), 24; "The Inflation Calculator" [Online resource], available at www.westegg.com/inflation/infl.cgi (accessed June 21, 2018).
57 Harron, "Eye in the Sky," 59–60.
58 These income figures were reported in Powers, "Photography at the Race Track," 31–32.
59 Harron, "Eye in the Sky," 60.
60 Robert Capon, "Improvements in or Relating to Recording Apparatus for Use in Investigating Speed," British Patent Office (November 10, 1927); "Photo Finish," *British Racehorse* (November 1950): 407.
61 Capon, "Improvements in or Relating to Recording Apparatus for Use in Investigating Speed," 407.
62 For more on Bing Crosby, see: *The South Bank Show: Bing Crosby* (Tele-Line, for BBC Television), Parts 1–3. Incidentally, after inventing a race-track strip camera, Del Riccio formed a company (Photochart) with Crosby to market the camera technology. However, when Del Riccio began making cameras for competing companies, Photochart (which still exists today, under the name Plusmic) sued Del Riccio for breaching the non-compete clause in his contract. Del Riccio claimed that he did not know that the contract still applied. To add insult to injury, Del Riccio then refused to pay his part of the legal fees to help Photochart fight off copyright abusers. Del Riccio and Photochart entered a court battle that lasted about a decade, and finally ended in 1954. As a result of these court decisions, Del Riccio was no longer allowed to claim royalties, and his extended contract and non-compete clause were both found to be valid: *Del Riccio vs. Photochart/Photochart vs. Del Riccio* (Civil Court – State of California 1946 (No. 15,293)/1949 (No. 17,115)/1952 (No. 19,362)/1954 (No. 19,855). Del Riccio later was hired by the United States Army Corps. of

Engineers to work with Captain (later, General) George Goddard to develop a form of strip camera for use in aerial photography. For more about the making of aerial photographs, see this manual: United States Corps of Engineers, *Engineering and Design: Photogrammetric Mapping, Manual* (Washington, DC: Department of the Army, 2002). Aerial strip photography was replaced in the 1970s by satellite mapping and imaging.

63 Anne-Marie Willis names Athol Shmith and B. Altson Pearl as the inventors of photo-finish technology: Anne-Marie Willis, *Picturing Australia* (Sydney: HarperCollins Publishers, 1989), 116.

64 Don Morehouse, "Tour of Aqueduct Race Track Photo Finish Booth, Interview by Author" (March 11, 2005). Race-track photo-finish technicians (who do not call themselves "photographers") use two cameras to record finishes, just in case one camera runs out of film or breaks. Since the late 1990s, most tracks have converted to semi-automatic digital technology, to reduce operator error and transmit photographs to judges sooner. A backup camera is still used.

65 "Did the Camera Lie?" *The Sydney Morning Herald*, www.smh.com.au/articles/2003/11/03/1067708135702html?from=storyrhs (accessed February 23, 2007); *ABC News*, "The Track," www.abc.net.au/tv/thetrack/ep4trans.htm (accessed March 23, 2007). Americans also eagerly accepted the photo finish. Don Morehouse – who has been making photo-finish images for race tracks for more than forty-five years for the New York Racing Commission – claims that photo finishes are not capable of inaccuracy, because the proper installation and operation of cameras is monitored closely by his employers, the state of New York's Racing Commission: Don Morehouse, "Tour of Aqueduct Race Track Photo Finish Booth, Interview by Author."

66 An elevated view of the finish line is advantageous because it allows the camera to see any horses which may be blocked by the horse in the foreground. To read more about experiments with tower heights and the possibility of suspending the camera on a cable directly over the finish line, see French Lane, "Improvements in Arlington Make It Ideal Track: 'Eye-in-the-Sky' and New Odds Board Features," *Chicago Daily Tribune*, June 22, 1936, 22; "Arlington Park Opens 30-Day Meet Tomorrow," *Chicago Sunday Tribune*, June 28, 193, Section 2, Page 4; Braven Dyer, "The Sports Parade: Column Conductor Finds Bing Crosby's Del Mar Plant Architectural Gem; Track Ready for Grand Opening," *Los Angeles Times*, 1935, A14. Incidentally, gambling has become even more profitable over time. In an average weekday in 2005, Aqueduct Race Track made about $12 million (US$15,354,196 in 2017) from on-site and off-track betting: Morehouse, "Tour of Aqueduct Race Track."

67 The average estimated speed of running horses on dry ground is 55 feet per second: Oswald Ziegler, ed., *Australian Photography* (Sydney: Ziegler Gotham Publications, 1947), 24.

68 A strip camera may record movement on the film from left to right, or vice versa. The direction of film's movement does not affect the veracity of the resulting image: Andrew Davidhazy, "Interview by Author" (2005). However, if racing objects are depicted running in one direction on film when they ran the other way in reality, patrons might lose faith in the photo finish's truth value. For this reason, race-track technicians print display photo-finish images of subjects moving in the "right" direction: Morehouse, "Tour of Aqueduct Race Track."

69 I have borrowed the phrase "smear of time" from Andrew Davidhazy, Professor in the Imaging and Photographic Technology department, School of Photographic Arts and Sciences, Rochester Institute of Technology. Interestingly, the words "time smear" also have been used to describe lenticular images – which also complicate any preconceived notions that there might be only *one* way to consider a two-dimensional image's relationship to time: K. Jaffey and S. M. Dutta, "Time Smear Corrected Multiplex Holographic Display of Computerized Tomography Data," Selected International Society for Optical Engineering Papers on CD-ROM, Vol. 12, Disc 1. Conference: "Stereoscopic Displays and Applications" (1983); Malgorzata Machniewska, "With the Wink of an Eye," *Afterimage*, Vol. 29, No. 6 (2002): 11.

70 For a step-by-step description of the making of a photo finish, see Kent Hollingsworth, "Fast Camera, Tick Marks Lay Wire on Horse's Nose in Photo Finish," *Lexington Leader*, April 22, 1959.
71 Jegerings, "Photography: Motion Effects Techniques," 19–22.
72 Morehouse, "Tour of Aqueduct Race Track."
73 Many thanks also go to Andrew Davidhazy and Jerome O'Neill for their assistance in my understanding of photo-finish technology.
74 "When Is a Photo Not a Photo," *Toronto Star*, August 23, 2004, Section D, Page 3; August 28, 2004, Section H, Page 6.
75 United States Rowing Club, "Digital Timing and Photofinish: What Can You Expect to See?" (2000), n. p.
76 Morehouse, "Tour of Aqueduct Race Track."
77 Barthes, *Camera Lucida*, 115.
78 Ibid, 89.
79 Charles Sanders Peirce, "Logic as Semiotic: The Theory of Signs," in *Semiotics: An Introductory Anthology*, Robert E. Innis, ed. (Bloomington: Indiana University Press, 1985), 1–23.
80 Joel Snyder and Neil Walsh Allen, "Photography, Vision, and Representation," *Critical Inquiry*, Vol. 2, No. 1 (Autumn 1975): 61–91.
81 Joel Snyder and Neil Walsh Allen specifically argue: "The mechanical relations which guarantee the validity of the photograph as an index of a certain kind of truth have been almost completely severed from the creation of visual likeness." Snyder and Allen, "Photography, Vision, and Representation," 159.
82 Snyder and Allen, "Photography, Vision, and Representation," 143–169. Interestingly, although Snyder and Allen describe the strip-photography photo finish in the text of this essay, the image they selected to illustrate their article is a rare hybrid made by two different processes. The contact print along the top edge of the photo finish featured divided, separate, sequential frames, and was made by an "Eye-in-the-Sky" video camera. These systems were quickly outmoded because they featured problematically slow shutter speeds of 165 frames per second and frequently failed to record an image of the first-place finish. The dominant image below the contact print reveals traces of "ghost legs," an artificial "finish line," streaked turf, spool imprints, and oval-shaped wheels. Those details are telltale signs of the strip-photography process. Sportsman's Park – or Hawthorne Park (Chicago, Illinois) – was unfortunate enough to have *just* purchased Harry I. Day's "Eye-in-the-Sky" video-camera system for $300,000 (about US$4.5 million in 2007) during the Great Depression in the summer of 1936, mere months before the strip-photography photo finish (developed by Lorenzo Del Riccio) was patented, sold to racetracks, and almost immediately became the industry standard. This image, a highly unusual hybrid of both techniques, suggests the race track's reluctance to discard the "Eye-in-the-Sky" system entirely. Rather, the "Eye-in-the-Sky" camera system at Sportsman's Park apparently was integrated with Del Riccio's strip-photography camera and a mirror system invented in 1946 by Athol Shmith and B. Altson Pearl in this undated photograph. While this image is not an exemplary sample of strip photography, it suggests that an awkward technological adolescence followed the introduction of photo-finish systems.
83 Snyder and Allen, "Photography, Vision, and Representation," 159.
84 As an extended-duration exposure, the photo finish finds a plethora of precedents, including the time-lapse images of the medium's earliest practitioners such as Niépce, Daguerre, and Talbot, the "chronophotography" of Étienne-Jules Marey, the Futurist "photodynamism" of Anton Giulio Bragaglia, Harold Edgerton's stroboscopic images, and the "scientific management" studies of Lillian and Frank Gilbreth, to name a few. A detailed examination of these numerous extended-exposure predecessors of the photo finish exceeds the capacity of this case study.
85 The word "panorama" is Greek in origin, and means "all seeing." The term was used in 1792 in the London *Times* to announce the exhibition of a 360-degree painting made by Robert Barker of a city view from an elevated vantage point and expanding to the distant horizon: Angela Miller, "The Panorama, the Cinema, and the Emergence of the Spectacular," *Wide Angle*, Vol. 18, No. 2 (1996): 34. Other scholarship includes this look at panoramic vision, which precedes photography's invention: Stephan Oettermann, *The*

Panorama: History of a Mass Medium, Deborah Lucas Schneider, trans. (New York: Zone Books, 1997). This exhibition catalog also explores the various panoramic techniques: Joachim Bonnemaison et al., *Collection Panoramas: Bonnemaison Photographies, 1850–1950* (Arles: L'Espace Van Gogh, 1989).

86 The history of the use of Kodak's Cirkut Camera is detailed in Robert B. MacKay, *America by the Yard: Cirkut Camera Images From the Early Twentieth Century*, first editiion (London: W. W. Norton, 2006), 8.

87 The Cirkut Camera was invented by William J. Johnston and patented in 1904. The Rochester Panoramic Camera Company manufactured it until the company was acquired by Eastman Kodak three years later. Kodak's aggressive marketing campaign was largely responsible for popularizing the Cirkut Camera: MacKay, *America by the Yard*, 10–11. Very few original Cirkut Camera negatives survive: "Once the class graduated, the convention adjourned, or the brigade disbanded, there was little chance of additional orders or incentive to find space to store bulky Cirkut images": MacKay, *America by the Yard*, 6.

88 Joachim Bonnemaison, Chart, Variations on Panoramic Image-Making, reproduced in Tania Ruiz Gutierrez, "Études sur le Temps et l'Espace dans l'Image en Mouvement: Tissage Vidéo, Objets Spatio-Temporels, Images Prédictives et Cinéma Infini" (Université Paris 1, 2004): 185.

89 Early twentieth-century Cirkut Cameras were marketed toward commercial photographers, although Kodak's advertisements worked to stir interest in the panoramic group photography by advertising the Cirkut to a mass audience. For more about Kodak's savvy marketing strategies, see Douglas Collins, *The Story of Kodak* (New York: H. N. Abrams, 1990); Dan Meinwald, "Picture Perfect: The Selling of the Kodak Image," *Frame/Work*, Vol. 1, No. 2 (1987).

90 Ralph Hyde, "Comments by Ralph Hyde on 'The Grand Panorama of London from the Thames'" (2000), available at www.motco.com/panoramas/001/hyde.asp (accessed April 12, 2008), n.p.

91 For more about the panorama's connection to a desire for extended vision, see Shelley Rice, "Boundless Horizons: The Panoramic Image," *Art in America* (December 1993): 58–71. According to Stephan Oettermann, enthusiasm for extended views of the landscape was sparked after the first unmanned hot-air-balloon flight in 1782 in France: Oettermann, *The Panorama*, 14.

92 Oettermann, *The Panorama*, 13.

93 Oettermann, *The Panorama*, 13.

94 Maarten Vanvolsem, "Photography: Seeing Time/Experiencing Time," in *Critical Realism in Contemporary Art: Around Allan Sekula's Photography*, Jan Baetens and Hilde van Gelder, eds. (Leuven: Leuven University Press, 2006), 79, 81.

95 Sigmund Freud, *Beyond the Pleasure Principle*, James Strachey, ed. and trans. (1920; reprint, New York: W. W. Norton, 1961), 20. Author's spelling has not been changed.

96 John H. Girdner, *Newyorkitis* (New York: 1901), 119.

97 Willy Hellpach, *Nervousness and Culture* (Berlin: 1902), 12.

98 Bernard Stiegler, *Technics and Time, 1: The Fault of Epimetheus* (Stanford: Stanford University Press, 1998), 15.

99 Bernard Stiegler, *Technics and Time*, 15.

100 Bernard Stiegler, *Technics and Time*, 15.

101 The public introductions of time-and-space-collapsing innovations, including the passenger train (1830), computer (1833), photography (1839), and the transatlantic telegraph (1844), occurred within a few years of each other. From the 1870s to 1940s, another surge in the introduction of time-and-space-collapsing inventions – including the photographic practices featured in this chapter's case study – occurred. This period brought the invention of the telephone (1876), automobile (1890s), cinema (1894), radio (1900–1910), the airplane (1903), and television (1939). It has been acknowledged as the longest period of sustained time-and-space-collapsing innovations, according to authors such as Stephen Kern: Stephen Kern, *The Culture of Time and Space 1880–1918* (Cambridge: Harvard University Press, 2003). This book argues that the phenomenon cannot be limited to just 1880–1918.

102 MacKay, *America by the Yard*, 9.
103 MacKay, *America by the Yard*, 17.
104 Miles Orvell, "Reproduction and 'The Real Thing': The Anxiety of Realism in the Age of Photography," in *The Technological Imagination: Theories and Fictions*, Teresa de Lauretis, Andreas Huyssen, and Kathleen Woodward, eds. (Madison: Coda Press Inc., 1980), 79.
105 Miles Orvell, "Reproduction and 'The Real Thing,'" 79.
106 "Double-enders" have a higher market value than other panoramic photographs: MacKay, *America by the Yard*, 15. Another relationship to temporality, the duration of Devereux's run, also is implied by his double presence in the image.
107 Oettermann, *The Panorama*, 21. The panorama also appeared at the beginning of a technological revolution that brought other innovations that eliminated other natural limits to human eyesight, including binoculars and microscopes, according to Oettermann, in *The Panorama*, 21; Merritt Ierley, *Wondrous Contrivances: Technology at the Threshold* (New York: Clarkson Potter Publishers, 2002), 19.
108 Although "peripheral photography" is the preferred term used by experts such as Andrew Davidhazy of the Rochester Institute of Technology, this form of photography goes by several other names. Mayanist art historians and archaeologists refer to the practice as "rollout photography." Europeans use the terms "roundshot photography" and "turntable photography." Davidhazy, "Camera for Conical Peripheral and Panoramic Photography"; Justin Kerr, "A Short History of Rollout Photography," *International Photographic Archive of Mayan Art* [Online journal], available at http://research.famsi.org/rollout.html (accessed February 13, 2007). Peripheral photography is also similar to aerial photography, in which the camera records the Earth's surface as the planet, the airplane, and the film in the camera all are moving. Like a photo finish, aerial photographs are images of elapsed time over a changing field of space. For more about the history of aerial photography, see Denis Cosgrove, "Contested Global Visions: One-World, Whole-Earth, and the Apollo Space Photographs," *Annals of the Association of American Geographers*, Vol. 84, No. 2 (1994): 279.
109 Curators and representatives of the British Museum do not have any additional information about their institution's role in the invention of the peripheral camera. Most sources cite this sixty-eight-year-old brochure as the "latest" research: Research Engineers Ltd., *Periphotography With the R. E. Periphery Camera* (London: Research Engineers Ltd., 1950), 1. The 1895 account of the peripheral camera's invention is reproduced in this 1950 report.
110 The Nebaj and Fenton vases are two of the most famous pottery works documented by the British Museum, using Hamilton-Smith's peripheral camera: Kerr, "A Short History of Rollout Photography," n.p. "Rollout" photographs of Mayan vases first were introduced on a wide scale to archaeologists and art historians in 1978. Until that time, all "rollout" images were hand-drawn by artists. Michael Coe's publication accompanying an exhibition of Mayan vases at the Princeton University Art Museum featured photographs by Justin Kerr. The publication of these peripheral photographs prompted a re-evaluation of the study of Mayan vase iconography: Kerr, "A Short History of Rollout Photography," n.p.

Bibliography

Adams, Gridley. "Letters to the Sports Editor: Defends Racing Camera." *New York Times*, July 11, 1936, 8.

Aldur, Martin. "The Representation of Movement in Art." *Burlington Magazine for Connoisseurs*, Vol. 23, No. 124 (1913): 204–207.

Aqueduct Race Track, Queens, NY. General information [Online track history]. www.nyra.com/Aqueduct/displaypages.asp?id=1&track=A (accessed March 11, 2005).

"Arlington Park Opens 30-Day Meet Tomorrow," *Chicago Sunday Tribune* (June 28, 1936): Section 2, Page 4.

Australian Government, Department of Intellectual Property [Online search engine] (accessed March 21, 2005).

Avery, Thomas Eugene. *Fundamentals of Remote Sensing and Airphoto Interpretation*. New York: Macmillan Publishing Company, 1992.

Bazin, André. "The Ontology of the Photographic Image." *Film Quarterly*, Vol. 14, No. 4 (1960): 4–9.

Belden-Adams, Kris. "A Wrinkle in Time." *Cabinet*, Vol. 25 (Spring 2007): 24–31.

Betts, John Rickards. "The Technological Revolution and the Rise of Sport." *Mississippi Valley Historical Review*, Vol. 40, No. 2 (1953): 231–256.

Bonnemaison, Joachim, et al. *Collection Panoramas: Bonnemaison Photographies, 1850–1950*. Arles: L'Espace Van Gogh, 1989.

Borchert, Donald M., and Stewart, David, eds. *Being Human in a Technological Age*. Athens: Ohio University Press, 1979.

Bovay, Jean-Pierre. "A Pledge for the Athlete." *Olympic Review*, Vol. 219 (1986): 44.

Bovay, Jean-Pierre. *Le Sens du Temps: Photografie et Perspective de Temps*. Bienne, Switzerland: Omega, Inc., 1988.

Bradshaw, C. M., ed. *Time and Behaviour: Psychological and Neurobehavioural Analyses*. London and New York: Elsevier Science B.V., 1997.

Braun, Marta. "The Expanded Present: Photographing the Moment." In *Beauty of Another Order: Photography in Science*, Ann Thomas, ed. New Haven: Yale University Press, 1997. 150–185.

Braun, Marta. *Picturing Time: The Work of Étienne-Jules Marey (1830–1904)*. Chicago and London: University of Chicago Press, 1992.

Braun, Marta. "Time and Photography." In *Tempus Fugit: Time Flies*, Jan Schall, ed. Kansas City, MO: Nelson-Atkins Museum of Art, 2000. 132–143.

Brookman, Philip. *Eadweard Muybridge*. London: Tate Publishing, 2010.

Brubaker, David. "Andre Bazin on Automatically Made Images." *Journal of Aesthetics and Art Criticism*, Vol. 51, No. 1 (1993): 59–67.

Burleson, Clyde W., Goldbeck, E. O., and Hickman, E. Jessica. *The Panoramic Photography of Eugene O. Goldbeck*. Austin: University of Texas Press, 1986.

"Camera Is Defended in Placing of Horses." *New York Times*, April 12, 1937, 24.

Capon, Robert. "Improvements in or Relating to Recording Apparatus for Use in Investigating Speed." British Patent Office, United Kingdom: November 10, 1927.

Cook, Jno. "The Pyrotechnics of Slit-Shutter Films." In *Light and Time* [Exhibition catalogue], Kathleen Kirka, ed. Chicago: Chicago Filmmakers, 1994. http://jnocook.net/cameras/film.htm (accessed June 20, 2018).

Crombie, Isobel. *Athol Shmith: Photographer*. Melbourne: National Gallery of Victoria, 1989.

Dagognet, François. *Étienne-Jules Marey: A Passion for the Trace*. New York: Zone Books, 1992.

Davidhazy, Andrew. "Synchroballistic Photography," "Instant Strip Photography," "Alternative Velocity Measurement Techniques in Synchroballistic Photography," "Camera for Photofinish Photography at Racetracks" [Online essays], n.d., available at http://people.rit.edu/andpph/articles.html (accessed Fall 2005).

Davidhazy, Andrew. "Simplified Streak/Strip and Scanning Photographic Systems." In *International Panoramic Photographers Association Conference*, Orlando, FL, 1985.

Davidhazy, Andrew. "Panoramic Cameras I've Made." Rochester, NY: Rochester Institute of Technology, 1987.

Davidhazy, Andrew. "Camera for Conical Peripheral and Panoramic Photography." *Current Developments in Optical Engineering and Commercial Optics*, Vol. 1168 (December 22, 1989: 1–10 doi:10.1117/12.962975

Davidhazy, Andrew. "A New Spin on Panoramic Photography." *Photo Educator International* (1992): 10–11.

Davidhazy, Andrew. "A Simplified Stroboscopic System for Motion Pattern Photography." *Tech Directions* (2000): 12–16.

Davidhazy, Andrew. "Development of a Novel Camera for Conical Panoramic Photography." Rochester Institute of Technology, Imaging and Photographic Technology Department, 2001, available from www.rit.edu/~andpph/text.strip (accessed March 18, 2007).

Davidhazy, Andrew. Interview by author and ongoing e-mail consultation. 2005–2018.

Davidhazy, Andrew. *The Basics of Strip Photography*. Rochester, NY: Rochester Institute of Technology, 2005.

De Duve, Thierry. "Time Exposure and Snapshot: The Photograph as Paradox." *October*, Vol. 5 (1978): 113–125.

"Del Riccio vs. Photochart." [Civil Court Cases Nos. 15293 (1946), 17115 (1949), 19362 (1952) and 19855 (1954)] State of California.

Delpire, Robert. "Avant-propos: Tout ou rien." In *Le temps d'un mouvement: Aventures et mésaventures de l'instant photographique*. Paris: Centre National de la Photographie, 1986.

Doane, Mary Ann. "Information, Crisis, Catastrophe." In *The Logics of Television: Essays in Cultural Criticism*, Patricia Mellencamp, ed. Bloomington: Indiana University Press, 1990.

Doane, Mary Ann. "Temporality, Storage, Legibility: Freud, Marey and the Cinema." *Critical Inquiry*, Vol. 22, No. 2 (1996): 313–343.

Doane, Mary Ann. *The Emergence of Cinematic Time: Modernity, Contingency, the Archive*. Cambridge, MA, and London: Harvard University Press, 2002.

Doane, Mary Ann. "Real Time: Instantaneity and the Photographic Imaginary." In *Stillness and Time: Photography and the Moving Image*, David Green and Joanna Lowry, eds. Brighton: Photoworks/Photoforum, 2006. 22–38.

Dyer, Braven. "The Sports Parade: Column Conductor Finds Bing Crosby's Del Mar Plant Architectural Gem; Track Ready for Grand Opening," *Los Angeles Times* (1935): A14.

Ebisch, Robert. "It's All in the Timing." *Sky Magazine* (December 1995).

Ellenbogen, Josh. "Photography and the Imperceptible: Bertillon, Galton, Marey." PhD diss., University of Chicago, 2005.

Ellenbogen, Josh. "Authority, Objectivity, Evidence: Scientific Photography in Victorian Britain." *Studies in the History and Philosophy of Science*, Vol. 39 (Spring 2008): 171–175.

Ellenbogen, Josh. "Camera and Mind." *Representations*, Vol. 101 (Winter 2008): 86–115.

"Faraway," "Explanation of a Riot." *The Blood-Horse* (August 21, 1937): 241.

"Finish Photography." *The Blood-Horse* (June 20, 1936): 743.

"French Photography." *The Blood-Horse* (July 10, 1937): 53.

Freud, Sigmund. *Beyond the Pleasure Principle*, James Strachey, ed. and trans. (1920; reprint, New York: W. W. Norton, 1961), 20.

Gehlen, Arnold. *Man in the Age of Technology*. New York: Columbia University Press, 1980.

Gere, Charlie. *Art, Time and Technology*. Oxford and New York: Berg, 2006.

Giedion, Sigfried. *Mechanization Takes Command: A Contribution to Anonymous History*. New York: Oxford University Press, 1948.

Gombrich, E. H. "Moment and Movement in Art." *Journal of the Warburg and Courtauld Institutes*, Vol. 27 (1964): 293–306.

Green, David. "Marking Time: Photography, Film and Temporalities of the Image." In *Stillness and Time: Photography and the Moving Image*. Brighton: Photoworks/Photoforum, 2006. 9–21.

Green, David, and Lowry, Joanna. *Stillness and Time: Photography and the Moving Image*. Brighton: Photoworks/Photoforum, 2006.

Haas, Robert Bartlett. *Muybridge: Man in Motion*. Berkeley, Los Angeles, and London: University of California Press, 1976.

Harron, Robert. "Eye in the Sky." *Collier's Illustrated Weekly*, September 19, 1936, 21, 59–60.

Hendricks, Gordon. *Eadweard Muybridge: The Father of the Motion Picture*. New York: Grossman Publishers, 1975.

Herzog, Amy. "Images of Thought and Acts of Creation: Deleuze, Bergson, and the Question of Cinema." *InVisible Culture*, Vol. 3 (2000), available at www.rochester.edu/in_visible_culture/issue3/herzog.htm

Hollingsworth, Kent. "Fast Camera, Tick Marks Lay Wire on Horse's Nose in Photo Finish." *Lexington Leader*, April 22, 1959.

"In the Wake of the News: Arlington Park." *Chicago Daily Tribune* (June 29, 1936): 19.

"The Instantaneous Process." *The Turf, Field and Farm*. June 18, 1886: 524.

"Interesting People." *American Magazine*, May, 1934, 48.

Ippolito, Jon. "Trusting Aesthetics to Prosthetics." *Art Journal*, Vol. 56, No. 3 (1997): 68–74.

"Is It a Push or a Pull?" *The Blood-Horse*, March 22, 1947: 833.

Jaffe, Irma B. "The Flying Gallop: East and West." *Art Bulletin*, Vol. 65, No. 2 (1983): 183.

Jegerings, Ron. "Photography: Motion Effects Techniques." *Darkroom & Creative Camera Techniques*, Vol. 14, No. 1 (1993): 20.

Jenkins, Reese. *Images and Enterprises: Technology and the American Photographic Industry, 1839 to 1925*. Baltimore: Johns Hopkins University Press, 1987.

Kern, Stephen. *The Culture of Time and Space, 1880–1918*. Cambridge: Harvard University Press, 1983.

Kern, Stephen. "Time and Art in Twentieth-Century Culture." In *Tempus Fugit: Time Flies*, Jan Schall, ed. Kansas City, MO: Nelson-Atkins Museum of Art, 2000. 22–34.

Kuipers, Dean and Aitken, Doug. *I Am a Bullet: Scenes from an Accelerating Culture*. New York: Crown Publishers, 2000.

Lawrence, Amy. "Counterfeit Motion: The Animated Films of Eadweard Muybridge." *Film Quarterly*, Vol. 57, No. 2 (2004): 15–25.

Lee, Gerald Stanley. *The Voice of the Machines*. New York: Mount Tom Press, 1906.

Lee, Pamela M. "Some Kinds of Duration: The Temporality of Drawing as a Process Art." In *Afterimage: Drawing through Process*, Cornelia H. Butler, ed. Cambridge, MA: MIT Press, 1999.

Lee, Pamela M. *Chronophobia: On Time and Art of the 1960s*. Cambridge, MA, and London: MIT Press, 2004.

MacDonnell, Kevin. *Eadweard Muybridge: The Man Who Invented the Moving Picture*. Boston and Toronto: Little, Brown and Company, 1972.

MacKay, Robert B. *America by the Yard: Cirkut Camera Images From the Early Twentieth Century*. London: W. W. Norton, 2006.

Marvin, Carolyn. *When Old Technologies Were New: Thinking About Electric Communication in the Late Nineteenth Century*. New York: Oxford University Press, 1988.

Marx, Leo. *The Machine in the Garden*. New York: Oxford University Press, 1964.

Massey, A. *The Independent Group: Modernism and Mass Culture in Britain, 1945–1959*. Manchester and New York: Manchester University Press, 1995.

Mayer, Ernest, Mayer, Frederic, and Pierson, Pierre-Louis. *La Photographie: considérée comme art et comme industrie*. Paris: Librairie L. Hachette et Cie., 1862.

Meers, Nick. *Stretch: The World of Panoramic Photography*. Mies: RotoVision, 2003.

Miller, Angela. "The Panorama, the Cinema, and the Emergence of the Spectacular." *Wide Angle*, Vol. 18, No. 2 (1996): 34.

Morehouse, Don. Tour of Aqueduct Race Track Photo Finish Booth in Action, and Consultation. March 11, 2005.

Morton, Stephen. "The Origins of the Panoramic Camera." *PhotoHistorian*, Vol. 104 (1994): 25–30.

Muybridge, Eadweard. *Scientific American*, Vol. 39, No. 19 (October 1878): 239.

Muybridge, Eadweard. "Improvement in the Method and Apparatus for Photographing Objects in Motion." United States Patent Office. United States of America, 1879.

Muybridge, Eadweard. "A Dead Heat." *Nature*, Vol. 26 (1882): 81.

Muybridge, Eadweard. *Animal Locomotion: The Muybridge Work at the University of Pennsylvania*. Philadelphia: J. B. Lippincott Co., 1888.

Muybridge, Eadweard. *Animals in Motion*. London: Chapman and Hall, 1899.

Norcross, Carl. "Don't Trust Your Eyes." *Collier's Illustrated Weekly*, September 17, 1938, 12, 27–30.

O'Neill, Jerome P. "The History of Race Photo Finish Photography: Interim Report of the Photo Finish Research Project" [Unpublished manuscript]. Rochester, NY: 1972.

Oettermann, Stephan. *The Panorama: History of a Mass Medium*. Deborah Lucas Schneider, trans. New York: Zone Books, 1997.

Orvell, Miles. "Reproduction and 'The Real Thing': The Anxiety of Realism in the Age of Photography." In *The Technological Imagination: Theories and Fictions*, Teresa de Lauretis, Andreas Huyssen, and Kathleen Woodward, eds. Madison: Coda Press Inc., 1980. 49–64.

Orvell, Miles. *The Real Thing: Imitation and Authenticity in American* Culture, *1880–1940*. Chapel Hill: University of North Carolina Press, 1989.

Parallel of Life and Art: Catalogue of the Exhibition. London: Institute of Contemporary Arts, 1953.

Perchick, Max. "Extending the Frontiers of Photographic Vision." *PSA Journal*, Vol. 62, No. 11 (1996): 26.

"Photo Finish." *British Racehorse* (November 1950): 407.

"Photo-Finish Contest to be Viewed in Court." *Los Angeles Times*, December 29, 1948, 7.

Pichel, Irving. "Stills in Motion." *Hollywood Quarterly*, Vol. 5, No. 1 (Autumn 1950): 8–13.

Portefin, Jules. "French Horse Painters of the Nineteenth Century." *British Racehorse* (November 1950): 395.

Powers, Ralph. "Photography at the Race Track." *Photo Technique* (June 1939): 30–33.

Prodger, Phillip. *Time Stands Still: Muybridge and the Instantaneous Photography Movement*. Oxford: Oxford University Press, 2003.

"Progressive Photography." *The Spirit of the Times*. October 11, 1879: 252.

Rabinbach, Anson. "Neurasthenia and Modernity." In *Incorporations*, Jonathan Crary and Sanford Kwinter, eds. New York: Zone, 1992. 178–189.

Rabinbach, Anson. *The Human Motor: Energy, Fatigue, and the Origins of Modernity*. Berkeley: University of California Press, 1990.

Research Engineers Ltd. *Periphotography With the R. E. Periphery Camera* [Report]. London: Research Engineers Ltd., 1950, 1–19.

Rice, Shelley. "Boundless Horizons: The Panoramic Image." *Art in America* (December 1993): 58–71.

Rodin, Auguste. *L'art: Entretiens reunis par Paul Gsell*. Paris: Bernard Grasset, 1924.

Ruiz Gutierrez, Tania. "Études sur le temps et l'espace dans l'image en mouvement. Tissage vidéo, objets spatio-temporels, images prédictives et cinéma infini." PhD diss., Université Paris 1, 2004.

Scharf, Aaron. "Painting, Photography, and the Image of Movement." *Burlington Magazine for Connoisseurs*, Vol. 104, No. 710 (May 1962): 186, 188–195.

Sheldon, James, and Reynolds, Jock. *Motion and Document, Sequence and Time: Eadweard Muybridge and Contemporary American Photography*. Andover, MA: Addison Gallery of American Art, 1991.

Smith, E. C. "Finish Photos." *Blood Horse* (February 22, 1936): 277.

Snow, C. P. *The Two Cultures and the Scientific Revolution*. New York and London: Cambridge University Press, 1993.

Snyder, Carl. "The World Beyond Our Senses." *Harper's Monthly Magazine* (June 1903): 117–120.

Snyder, Joel, and Allen, Neil Walsh. "Photography, Vision and Representation." In *Reading into Photography: Selected Essays, 1959–1980*, Shelley Armitage, William E.

Tydeman, and Thomas Barrow, eds. Albuquerque: University of New Mexico Press, 1982. 61–91.

Solnit, Rebecca. *River of Shadows: Eadweard Muybridge and the Technological Wild West.* Middlesex, England: Viking Penguin, 2003.

Souriau, Paul. *The Aesthetics of Movement.* Amherst: University of Massachusetts Press, 1983.

The Spirit of the Times (June 3, 1882): 513.

The Spirit of the Times (June 26, 1886): 702.

Sternberger, Dolf. *Panorama of the Nineteenth Century* (1955), Erich Heller, trans. New York: Mole Books, 1977.

Stiegler, Bernard. *Technics and Time, 1: The Fault of Epimetheus.* Stanford: Stanford University Press, 1998.

Sutton, Damien. *Photography, Cinema, Memory: The Crystal Image of Time.* Minneapolis: University of Minnesota Press, 2009.

Sutton, Damien. "Delay. Notes on Photography as Non-Representational Thinking." In *Time Expanded.* Madrid: La Fábrica/PHotoEspaña, 2010. 145–151.

Swinton, A. A. C. "Photographing the Unseen." *Cornhill Magazine* (January 1896): 290–296.

Thompson, E. P. "Time, Work-Discipline, and Industrial Capitalism." *Past and Present*, No. 38 (1967): 56–97.

United States Patent and Trademark Office. Online Database of Issued. Available from http://patft.uspto.gov/netahtml/search-bool.html (accessed March 22 2005).

University of Pennsylvania. *Animal Locomotion: The Muybridge Work at the University of Pennsylvania.* Philadelphia: J. B. Lippincott Co., 1888.

Vanvolsem, Maarten. "Photography: Seeing Time/Experiencing Time." In *Critical Realism in Contemporary Art: Around Allan Sekula's Photography*, Jan Baetens and Hilde van Gelder, eds. Leuven: Leuven University Press, 2006. 79, 81.

Vanvolsem, Maarten. *The Art of Strip Photography: Making Still Images With a Moving Camera.* Leuven: Leuven University Press, 2011.

Watts, Jennifer A., and Bohn-Spector, Claudia. *The Great Wide Open: Panoramic Photographs of the American West.* London: Marrell Publications Limited, 2001.

"When Is a Photo Not a Photo?" *Toronto Star*, August 23 and 28, 2004, 3D and 6H.

Willis, Anne-Marie. *Picturing Australia.* Sydney: HarperCollins Publishers, 1989.

Ziegler, Oswald, ed. *Australian Photography.* Sydney: Ziegler Gotham Publications, 1947.

5 A "Tapestry" of Synthetic, Hypothetical Digital Time

NASA's Whole-Earth Photographs as "Data Visualizations"

After taking 100 micrograms of LSD one fateful evening in February 1966, Stewart Brand stood on his San Francisco rooftop and had a revelation that launched what would become an obsessive quest.[1] He suddenly realized that a color photograph of the entire Earth, basking in full sunlight, taken from outer space, had never been made before. Brand felt such an image might have the capacity to transform the way everyone felt about the planet, and could bolster support for one of his beloved causes, the environmental movement.[2]

When he woke up the next day, Brand ordered hundreds of buttons and posters bearing the slogan "Why haven't we seen a photograph of the Whole Earth yet?" and mailed them to the National Aeronautics and Space Administration (NASA), to members of the United States Congress, to diplomats, officials from the United Nations, and to Soviet scientists.[3] Brand also sold buttons for 25 cents at the University of California at Berkeley. He was thrown off campus, but still managed to sell out of buttons at Stanford University, Columbia University, Harvard University, and at the Massachusetts Institute of Technology.[4] One man's revelation, born on a rooftop in San Francisco, soon sparked a grassroots campaign of international proportions. (It should be noted that, in 1948, British astronomer Fred Hoyle suggested that a photograph of the full Earth, as it is seen from outer space, could unleash "a new idea as powerful as any in history will be let loose."[5] While Brand was not the first to have this idea, his grassroots campaign calling for photographs of the Earth gained mass-cultural traction.)

However, it would be six years until Brand's demands for a *color* photograph revealing the *entire* Earth basking in full sunlight, made by a *human being*, was fulfilled. Months after Brand's campaign started, NASA's Applications Technology Satellite-I (ATS-I) weather satellite transmitted a blurry set of black-and-white images of the full planet, taken by the satellite, from the purview of 23,000 miles above its surface. And in November of 1967 – one year after Brand's campaign began – another unmanned craft, the ATS-III, photographed the Earth with color film from a distance of 22,300 miles above the surface, in order to observe air-mass motion, cloud heights, rainfall, pollution, and natural disasters. The satellite's Multi Color Spin Scan Camera (MCSSC) camera automatically snapped photographs every half-hour, for twenty-four hours (Fig. 5.1). One of the resulting images represented the entire Earth in full sunlight, with a prominent view of South America. Swirls of cloud cover represent a planet in the throes of dynamic meteorological change. Strong storms wreak havoc on the Great Lakes and Eastern Canada, and dense cloud cover approaches the western edge of South America. Thick cloud cover coalesces at both poles. Land masses enjoy varying hues of

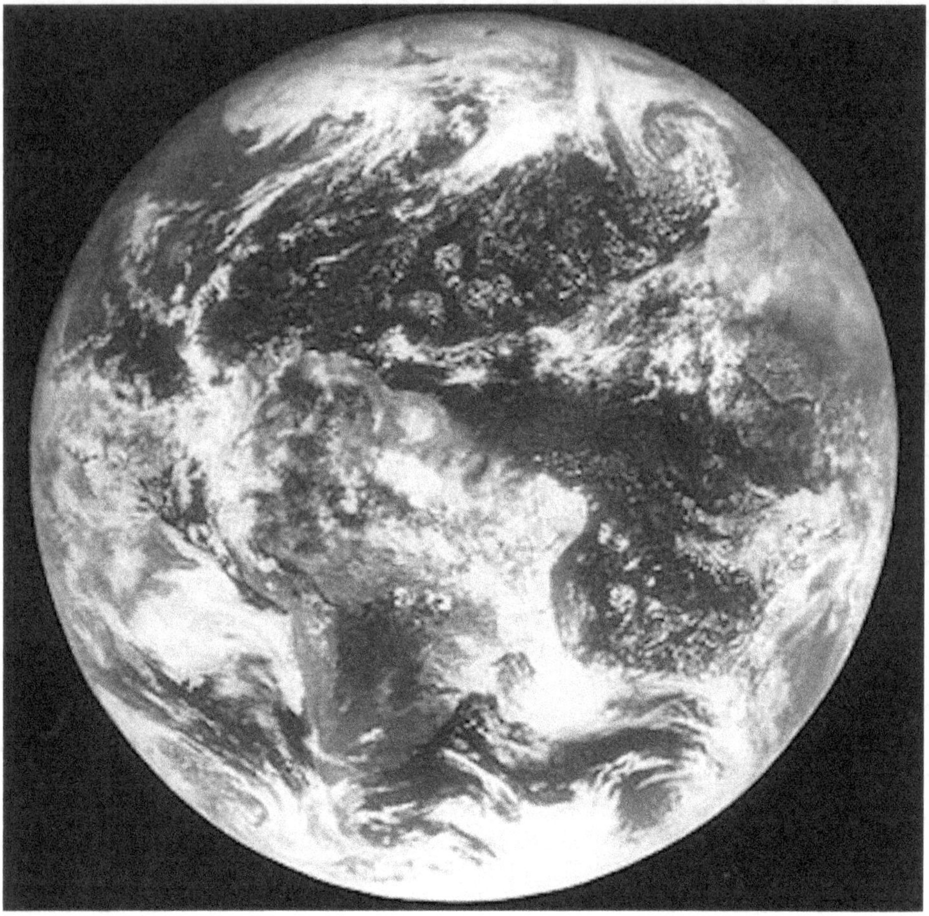

Figure 5.1 National Aeronautics and Space Administration, *ATS-III Mission Photograph*, November 18, 1967. Courtesy of the National Aeronautics and Space Administration.

chocolate brown, and bodies of water in the center of the Earth appear black in the ATS-III photograph. The edge of the bottom half of the planet is more illuminated than its counterpart at the top of the Earth, suggesting differential atmospheric light refraction.

The ATS-III photographs were relatively blurry, but – unlike the ATS-I images – were published in newspapers and magazines worldwide. In the fall of 1968, Brand compiled the photographs from the ATS-III mission and published them as a book titled *Whole Earth Catalog*. It sold well, and generated additional public interest and support for NASA's missions.[6] This, in turn, encouraged NASA to document their explorations of outer space with photography, in order to hold public support and gain additional government funding. Still, Brand lamented that the ATS-III photographs were lacking, because they were not snapped by the human hand, but by an unmanned craft. This is to say, it mattered to Brand that human eyes had not yet experienced and photographed the scene first-hand.

However, on December 22, 1968, astronaut William Anders from the Apollo 8 mission photographed the Earth in color, from a distance of 240,000 miles away (Fig. 5.2).[7] The image, titled *Earthrise (View of Rising Earth About Five Degrees Above the Lunar Horizon)*, revealed the planet in partial darkness. The Apollo 8 crew did not notice this "Earthrise" on their first three orbits around the moon because they were busy with other tasks. Although one of the Apollo 8 astronauts' missions included taking photographs of potential lunar landing sites and other views of scientific interest, *Earthrise* resulted from an inspired impulse to photograph, rather than a scheduled assignment. The on-board voice recorder captured the following excited conversation among the three astronauts, testifying to the effects of first-hand viewing of the Earth from outer space that Brand desired – even if the entire planet was not in view:

COMMANDER FRANK BORMAN: Oh my God! Look at that picture over there! Here's the Earth coming up. Wow, that is pretty!
ANDERS: Hey, don't take that, it's not scheduled.
BORMAN: (Laughter) You got color film, Jim?
ANDERS: Hand me that roll of color quick, will you.
JAMES LOVELL: Oh man, that's great!

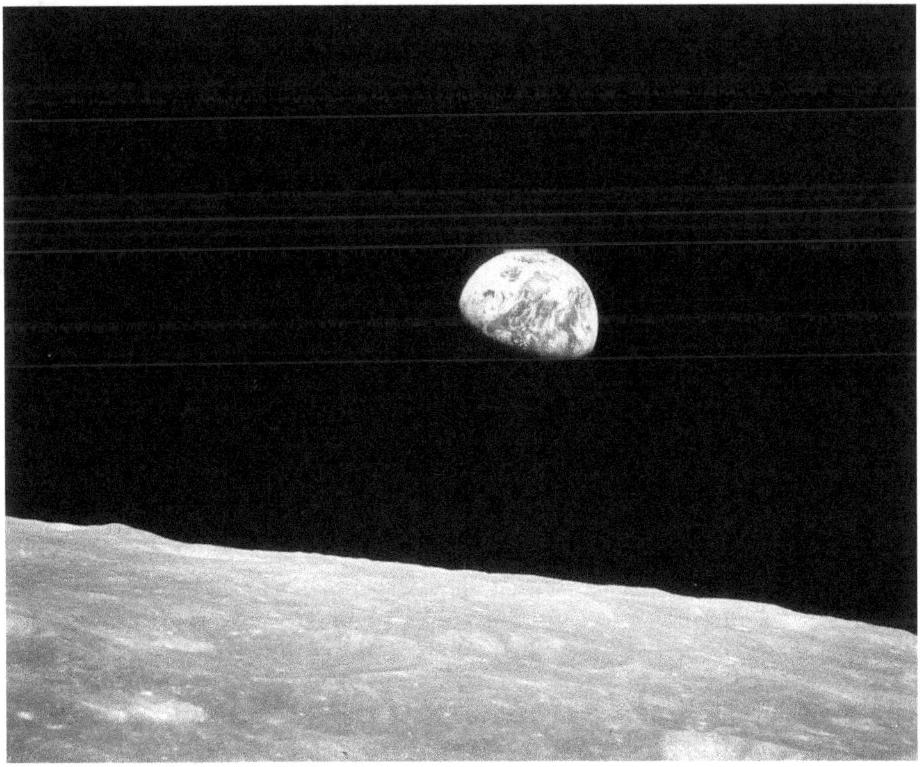

Figure 5.2 William Anders, *Earthrise*, Apollo 8, December 24, 1968. Courtesy of the National Aeronautics and Space Administration/Public Domain.

ANDERS: Hurry. Quick.
LOVELL: Take several of them! Here, give it to me ...
BORMAN: Calm down, Lovell.[8]

In hindsight, Borman later recalled in his autobiography: "It was the most beautiful, heart-catching sight of my life."[9] As Anders himself noted, the Earth "should be thought of as [a] fragile Christmas-tree ball which we should handle with considerable care."[10]

Earthrise depicts the planet Earth in the center of the frame, with its land masses obscured in swirls of cloud covering. The gray, crater-marked surface of the Moon fills the right edge of the photograph, revealing the point of view from which the image was made – and a reversal of our usual view of the Earth's primacy. *Earthrise* humbly represents our known world as a small, delicate, isolated, planet dwarfed by the vast darkness of outer space. Poet Archibald MacLeish wrote of *Earthrise*: "Earth was no longer the World, but a small, wet, spinning planet in the solar system of a mirror star off at the edge of an inconsiderable galaxy in the immeasurable distances of space."[11] According to Denis Cosgrove:

> MacLeish's world dramatically underlined Modernist beliefs in the mimetic truth of photography, the power of the synoptic gaze (which his language clearly regarded as gendered male), and equally, one-worldism, that is, the claims of American missionary ideology, of which this image stands as empirical proof.[12]

The photograph evokes viewers' knowledge of both the *known* and the *unknown* as humankind's home is revealed as a diminutive component of a vast, fairly unexplored, universe. *Earthrise* rhetorically fuels positivist advance and exploration by illustrating the next frontier of American technological and scientific conquest. An appetite for seeing the Earth from new vantage points was inextricably bound to a broader cultural enthusiasm for extended vision encouraged by various revolutionary ruptures in perceptions of space and time in science and technology.

Temporally, the photographic medium lends viewers the opportunity to enjoy pictures of multiple spaces and times through the camera's surrogate eye. Absent viewers enjoy a delayed virtual presence as observers to subjects across the space/time continuum. Viewers also may review photographs at other points in time, further multiplying and confounding this binary play of temporal absence and presence. Based on the speed of light, the instantaneous-time photograph of the Earth represented light that was about a second "old" at the time of its recording on camera. It marks a departure from the "fresh," split-second immediacy of light travel from an object to a camera that we have come to expect of the fraction-of-a-second exposure. *Earthrise* makes multiple references to the past by also implying the variable duration of the speed of light's differential travel from the atmospheres/surfaces of the Earth and Moon to the film inside of the camera aboard the Apollo 8 spacecraft. Light refracted from the Moon is less deferred than that from the Earth. Therefore, *Earthrise* represents a generalized past of a Barthesian *that-has-been*, while also to the deferred past of the Moon's reflected light, and the even-more-deferred past of the Earth's reflected illumination.

Earthrise also represents two celestial bodies in the throes of their respective and ongoing temporal acts of rotation and orbit. Upon viewing *Earthrise*, Braun commented that "this photograph is a picture of movements – the diurnal rotation of the earth and the much longer rotation of the moon," despite the fact that this movement is "too slow" to find representation on the camera's film.[13] Braun suggests that these processes are nonetheless implied in *Earthrise*. The simultaneously operating, cyclical planetary movements define seasons, nights/days, and years – the very parameters by which humans measure time. Swirls of clouds in the photograph imply that Earth also is dynamically engaged in various changeable processes of meteorological phenomena that are affected by seasonal cycles. The surface of the Moon, in comparison, is not subject to the same degree of apparent meteorological dynamism and change. The dynamic, climate-event-wracked Earth, is prone to incessant change, and is intrinsically and intensely temporal.

The still photograph *Earthrise* represents an isolated fragment of time that is abstract and estranged from our experience of Earth's planetary time as an ongoing set of processes in a continuum. For instance, the repetitive process of the *Earthrise* – and its inevitable antithesis, Earth-"sunset" – comprise our familiar concept of the "day," and are thus intrinsic to our very conception of time itself. But while the "day" is a known quantity to us, it differs from the length of a "day" on the Moon, and from all known planets in our solar system. Humankind's temporal concept of the "day" is therefore a relative concept. Like the experience of time itself, photography's fragmentary views of time are equally abstract.

Even after NASA's publication of the ATS-III and *Earthrise* photographs, Brand's grassroots campaign for a color photograph representing the entire Earth in full sunlight – taken by a human being – remained unfulfilled.[14] During its coast toward the Moon, one of three astronauts aboard the Apollo 17 spaceship, Jack Schmitt, shot a sequence of eleven color photographs of the Earth with a hand-held Hasselblad camera (which later broke, damaging most of the other photographs).[15] At 5:33 a.m., Eastern Standard Time, on December 7, 1972, an astronaut captured image AS17-148–22,727, at a distance of about 21,750 nautical miles from the Earth's surface (Fig. 5.3). In this photograph, the Earth appears, center-frame, and basking in full sunlight. (The only "imperfection," as some North American commentators saw it, was the fact that Africa and Antarctica, not the United States of America, were the points of focus.)[16] The Sahara Desert appears cloud-free, but swirls of clouds covering Antarctica imply dynamic time and motion.[17] The continent of Africa enjoyed exposure to daylight sunshine, and because the December solstice was imminent, Antarctica also was illuminated.[18] The photograph conveys seasonal and astrological temporal specificity. NASA's official caption for *Big Blue Marble* describes it as follows:

> View of Earth as seen by the Apollo 17 crew traveling toward the moon. This translunar coast photograph extends from the Mediterranean Sea area to the Antarctica south polar ice cap. This is the first time the Apollo trajectory made it possible to photograph the south polar ice cap. Note the heavy cloud cover in the Southern Hemisphere. Almost the entire coastline of Africa is clearly visible. The Arabian Peninsula can be seen at the northeastern edge of Africa. The large island off the east coast of Africa is the Republic of Madagascar. The Asian mainland is on the horizon.[19]

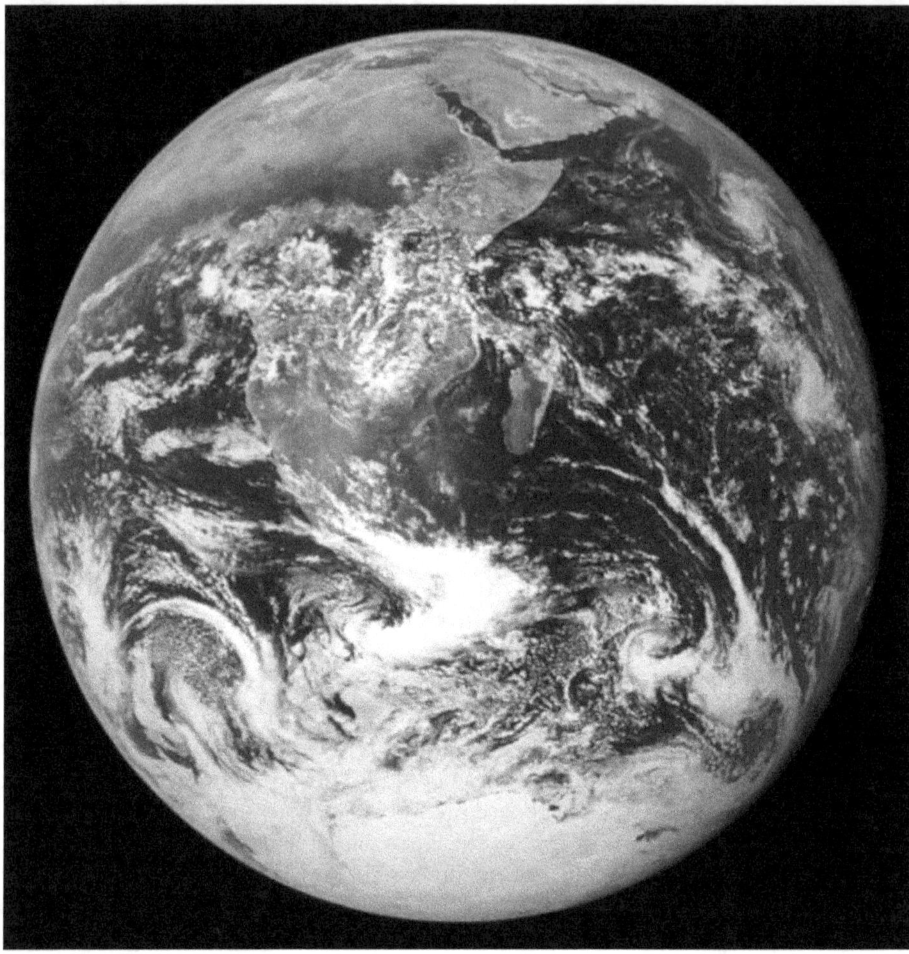

Figure 5.3 Jack Schmitt, *Big Blue Marble* #22,727, Apollo 17, 1972. Courtesy of the National Aeronautics and Space Administration/Public Domain.

According to NASA, the image, more commonly known as *Big Blue Marble*, has a specific "narrative time" context, within the Apollo 17 mission. It was made in a fraction-of-a-second exposure time with an unspecified duration that occurred at five hours and six minutes after launch, and about one hour and forty-eight minutes after the spacecraft left its parking orbit around the Earth to begin its trajectory to the Moon.[20] That is to say, NASA's documentation for *Big Blue Marble* is temporally specific to the minute, yet not to the second – or fractional second. Astronauts participated in flight simulations in advance of the mission, in part to learn how to take photographs that compensated for spacecraft roll, limited sight-lines, odd angles, sun coverage, and planetary rise and set times. Windows for taking the Apollo 17 photographs were computed in advance. According to Cosgrove:

> The extremely detailed plans for photographic procedures left little room for initiative on the part of the astronauts. Among the major constraints were: exposure times in the absence of light-diffusing atmosphere; insufficient time to determine optimal settings and vantage points; and weight limits on the number and variety of lenses and film.[21]

While *Big Blue Marble* exists within the "narrative time" context of this particular Apollo 17 mission, it also is positioned within the broader cultural phenomenon of the "space race," and within a narrative of these astronauts' experience of the morning of December 7, 1972. As Chapter 3 of this book has suggested, this dimension of the "narrative time" contexts of a photograph are mutable, multifarious, subjective, creative, and varies from viewer to viewer. And as it has been historicized in this chapter, the Apollo 17 *Big Blue Marble* photograph also inspires Brand's San Francisco-centered U.S. counter-culture, pro-environment movement in the mid- to late 1960s.

Moreover, in order to publicize and garner public support for the Apollo missions, NASA placed *Big Blue Marble* in the public domain, so that it would be available at no charge to the media and the public.[22] The original photograph was rotated to align the top of the image with the Earth's ordinal direction of "north," in compliance with mapping conventions. The image also was cropped to center the Earth in the frame before distribution.[23] NASA representatives have speculated that *Big Blue Marble* has become the most widely distributed image in history.[24] Thus, the photograph's potential, plural interpretive "narrative times" are especially rich with possibility.

In *Big Blue Marble*, time is represented in several other ways, too. Like *Earthrise*, the light refracted from the Earth reached the astronauts' camera approximately one second later to make *Big Blue Marble*. Thus, this image does not represent its subject's presence in a conventional past tense – or, in Barthes's terms, a *that-has-been* – but it also implies a *second-deferred* past tense. The delay caused by time's travel is virtually undetectable in the photographing of objects that are positioned just *yards* away – as opposed to thousands of miles, or even *light-years* away – from the camera. The photograph thus makes implicit reference to a duration of time during which light traveled from the planet to the camera's light-sensitive surface, a temporal duration that typically escapes our attention, but is inherent in all photographs of the Earth.

Thus, in *Big Blue Marble*, time is extended. What we see is *a duration* of time. Yet the image was snapped in an instantaneous fraction of a second. These apparently conflicting attributes of time – that is, time, as it appears in instantaneous photography – were explained by Charles Sanders Peirce: "Even what is called an 'instantaneous photograph,' taken with a camera, is a composite of the effects of intervals of exposure more numerous by far than the sands of the sea."[25]

But the temporal attributes of *Big Blue Marble* are yet more complicated. Light collected on film was actually reflected from the Earth's surface in a split second in the *past* that took time to travel through space and collect on film.[26] Not only does the image reveal the conventional appearance of things in the past – as is conventionally assumed of instantaneous photographs – it also presents the appearance of the Earth in a time *previous* to the past-time assumed by the viewer. NASA's *Big Blue Marble* then represents an image with an *instantaneous* relationship to time (thanks to the camera's quick shutter operations), but which actually depicts a *duration* of time and movement that transpired in the *past*. The photograph exemplifies Geoffrey Batchen's suggestion that photography is "*in* time but not *of* time."[27]

Big Blue Marble's *Re-Generation*: Digital Whole-Earth Image-Making Practices

Despite a flurry of excitement over these technological milestones, funding and interest in America's space program began to wane in 1972, as Americans' attentions instead turned to the country's involvement in the Vietnam War.[28] As a result, Apollo 17, the spacecraft from which *Big Blue Marble* was photographed in 1972, was *the only manned spacecraft ever* to orbit far enough away from the Earth that it could view the entire planet.[29] Later missions (including Space Shuttles, Soyuz and International Space Station, and various satellites) have flown too close to the Earth to observe the globe in its entirety.

Therefore, according to William Stefanov, Senior Geoscientist for NASA's Image Science and Analysis Laboratory, any image representing the entire Earth in full sunlight created after 1972 is a composite photograph. Starting in the 1990s, NASA has created digitally manipulated whole-Earth images, which, since 2000, have been titled *Big Blue Marble: Next Generation*, in tribute to the Apollo 17 mission.[30]

These images are made with the aid of computer software, which blends photographs taken over a duration of time from dozens of satellite images to produce an image of the Earth in full sunlight. *Big Blue Marble: Next Generation* digital composite images are difficult to distinguish from unaltered photographs from space, such as *Big Blue Marble*. According to the website of one of the producers of whole-Earth digital-manipulation software, ARC Science Simulations, the goal of the programs is to produce "photo-realistic," "accurate, high quality earth imagery ... [that] can compare remarkably well to photographs from space."[31] ARC also offers clients images photo-realistic simulated images of Mars, Jupiter, the Moon, and of the Paleolithic Earth at 30- to 40-million-year intervals, dating as much as 600 million years ago.[32]

The process of making whole-Earth photographs is arduous and expensive.[33] ARC Science Simulations' popular composite-generating software, for instance, connects computer operators to dozens of steady streams of instantaneous, photographs of the surface of the Earth that are taken continuously by the company's satellites, which are programmed to photograph every part of the Earth's surface, without cloud covering, at least one time.[34] For instance, in 1992, *each* Landsat satellite orbited the Earth fourteen times each day.[35] Daylight imaging passes are southbound ("descending"), northbound passes cover the dark side of the Earth ("ascending"). In 1992, it took 252 orbits – which transpired every eighteen days – for each satellite to cover the entire Earth. In other words, a singular satellite passes over the same point on the Earth's surface once every eighteen days. This time has gradually been reduced by the addition of more satellites. Because dozens of satellites are orbiting the Earth at any given time, sections of the planet are photographed multiple times. This satellite data is accurate to a resolution of 500 meters, or about 1,640 feet, per pixel.[36] NASA's Image Science and Analysis Laboratory employed a customized version of ARC's system and supplements that data with additional information from NASA's eighteen satellites.

Data collected by various satellites are transmitted to ground receiving stations for processing, labeling, correction, and storage as binary code.[37] When the software determines that the various satellite cameras have gathered sufficient information to create an image of the mathematical "probability of appearance" of details on and above the planet's surface, the digital compositing process begins. The data-collection span of time is subjective, and is related to both the available satellite technology and weather conditions. The image from 1996 (Fig. 5.4), which will be discussed in detail shortly, was made from *four months* of data. The whole-Earth photograph from the year 2002 (Fig. 5.5), in comparison, was made from *one day* of data collected on February 8. The culture of temporal investment at NASA encouraged ever-decreasing required data-collection windows, in favor of producing a

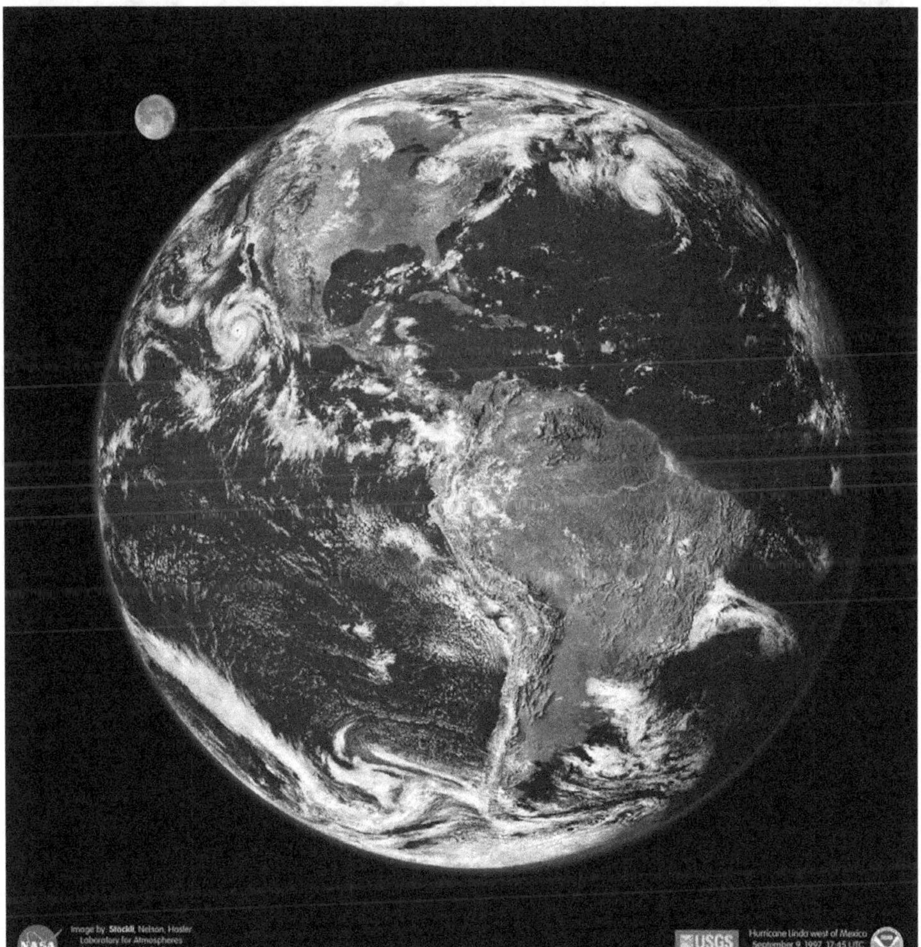

Figure 5.4 Reto Stockli and Alan Nelson, *Earth – The Big Blue Marble, Made by Data From a Geostationary Operational Environmental Satellite*, 1996. Courtesy of the National Aeronautics and Space Administration/Public Domain.

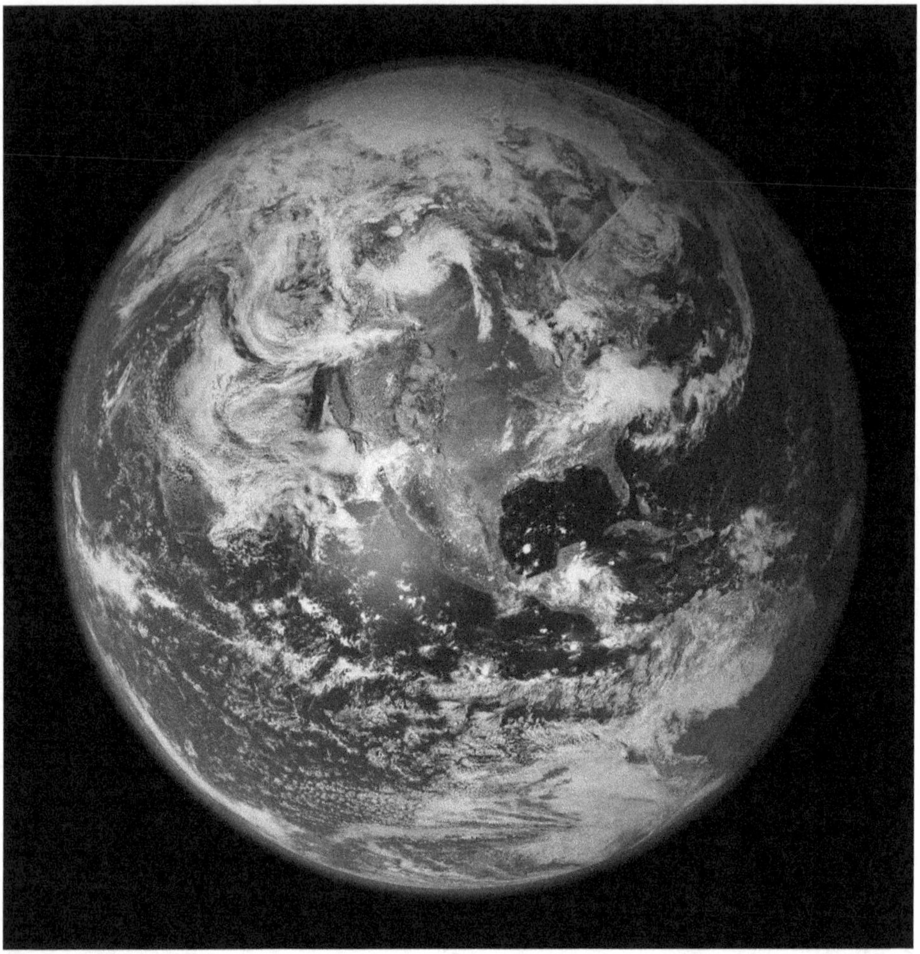

Figure 5.5 NASA Goddard Space Flight Center Image by Reto Stöckli (land surface, shallow water, clouds). Enhancements by Robert Simmon (ocean color, compositing, 3D globes, animation). *The Blue Marble: Next Generation, West Global View*, Terra-MODIS sensor/data. Date of data: February 8, 2002. Visualization date: February 8, 2002. Courtesy of the National Aeronautics and Space Administration/Public Domain.

photograph for online posting to the public-accessible Visible Earth website as quickly as possible.[38]

To make a whole-Earth digital image, the numerical visual information from multiple photographs is "averaged" to determine the most likely appearance of that parcel of the Earth over a technologically determined duration of time. If clouds covered most of America during the period of data collection, the country is likely to be obscured by clouds in the resulting image. Weather is thus an important variable during data collection.

At this point, surface-appearance data are reshaped, mathematically and perspectivally, and are instructed to adhere to the spherical, three-dimensional output model of the entire globe. In this process, photographs of half-second-deferred light captured by satellites much closer to the planet is represented as second-deferred light from at least 20,000 nautical miles from the Earth's surface. Any skewed photographing angles are "corrected."

However, operators can choose to override the computer-generated "average-appearance" cloud cover produced by the data set and mathematical figures – also called the computer's "decision trees."[39] They can opt instead to make a thinly clouded, or even cloud-free, picture of the Earth.[40] Because an organization such as NASA uses its images as points of visual comparison to track phenomena such as erosion, to monitor the location and damage caused by forest fires, and to study the effects of global warming, it is essential for their visual representations to appear plausibly realistic. Therefore, NASA's Image Science and Analysis Laboratory technicians claim to avoid overriding the mathematical "decision trees," unless an element like clouds, for instance, might obscure the view of the desired surface phenomenon.[41] Modifications also are made to the data gleaned from the dozens of satellites, according to NASA technician Reto Stöckli, to correct the "cloudy pixels, missing time periods, and processing artifacts [that] can disturb satellite-derived data" by "compositing based on empirical filters . . . to remove such irregularities."[42] After satellite data is "cleaned," reshaped, mathematically "averaged," and subjected to operators' aesthetic alterations, a digital whole-Earth composite photograph is generated. While this process sounds simple, according to Stöckli,

> The creation of seamless cloud-free spatial and temporal composites of Earth's surface is, however, not a trivial task. It is dependent on sophisticated atmospheric corrections and cloud screening. Even then, cloudy pixels, missing time periods, and processing artifacts can disturb satellite-derived data. Temporal compositing based on empirical filters . . . can be used to remove such irregularities.[43]

These "temporal compositing" filters make spatial-temporal adjustments to enable the flow of non-synchronously collected data samples appear to have been gathered synchronously, by evening out the varying light quality of the collected data. These filters also give the entire finished composite image the appearance of having been made under uniform lighting conditions, at a fraction of a *single* moment in time.

The digital whole-Earth photograph titled *The New World: GOES (Geostationary-Orbiting Environmental Satellite) Clouds Over the Face of the Earth, August 2, 1997* (Fig. 5.4) is a composite image that appears very similar in hue saturation, cloud covering and degree of surface detail to the 1972 *Big Blue Marble* single-exposure photograph – likely because the original was its color-palette model.[44] Although the digital image could be mistaken as a single-exposure image, the 1997 composite photograph has a different relationship to time. It was made from four months of cumulative data, in order to assure that every part of the Earth's surface eventually was visible below the cloud cover. This surface detail has a spatial resolution, or level of detail, of 1 square kilometer per pixel. In this photograph, North and South America, not Africa, are in central focus. Swirls of dynamic clouds avoid obscuring the western half of the United States, but are not omitted altogether, because a cloudless country might suggest that blatant alteration of the mathematical probability of the appearance of the surface data did take place.

The *Big Blue Marble* is a photographic "data visualization" of the Americas, as they would have appeared in "average" appearance, over four months – with undisclosed aesthetic modifications of this data. The resulting whole-Earth image corresponds to an abstract, hypothetical, imagined time. The planet never existed in a time-space continuum as it appears in a whole-Earth digital-composite photograph.

With respect to representing time, the *Big Blue Marble*, like all composite whole-Earth images, provides a departure from temporality's lived experience. At first glance, this digital image functions much like the single-exposure images *Earthrise* and *Big Blue Marble*. Reflected light that left the Earth's surface approximately a half-second earlier is collected as an instantaneous photograph. The image's "*that-has-been*" is an additional half-second deferred, and this condition makes reference to the physical act of light's travel over a duration of time. However, in the digital composite image the *Big Blue Marble*, this representation of time is rendered yet more complex. Rather than resulting from the travel of reflected light during *one* fraction-of-a-second camera exposure from a single vantage point, the image is the product of a *four-month-long duration* of taking billions of fraction-of-a-second photographs of the Earth' surface from the point of view of more than two dozen satellites. Computer synthesis of this data also required a significant and variable amount of time to convert satellite data into a representation of a spherical Earth. The process of making a whole-Earth image also implies an unseen temporal duration.

In *The New World*, a singular, fraction-of-a-second photograph is *not* selected as "the" image of a 500-meter "pixel" that represents that portion of the Earth's surface. Instead, this bit of visual data is translated into binary code, blended and averaged with *other* collected data that represent this part of the planet to calculate the "usual" appearance of this 500-meter "pixel" of the Earth. This practice of image-making is computational in nature, and visual/aesthetic in effect. In other words, what we see in *The New World* is a digital synthesis of a hypothetical, imagined image of how the Earth might appear from outer space over a (plural) durée of elapsed time. Thus, whole-Earth photographs offer an imaginary view of elapsed, multi-layered, multiplicitous times. By making such a view of time possible, these digital composites, in turn, introduce a synthetic temporality that enables viewers to inhabit multiple times, simultaneously, by compressing them into a lasting image.

Nicholas Mirzoeff suggests that digital photography has collapsed processing times and made the circulation of images more immediate.[45] Indeed, this compression of time pervades darkroom-free, digitally distributed whole-Earth digital composite practices. But it accelerated as the digital processes themselves evolved. For example, *Big Blue Marble: Next Generation* (Fig. 5.5) resulted from millions of photographs collected by MODIS sensors aboard the Terra satellites – which have been used for data collection since the year 2000 – from *one day* of data. This image reveals a greater degree of surface detail than *The New World*, according to Stefanov:

> Coupled with improved geolocation data, computer hardware, server connectivity, and image processing made it become possible to produce and serve frequently

> updated, global mosaics at approximately one kilometer per pixel or five kilometers per pixel resolution.[46]

Advances in data processing also improved image clarity, and gave "highly reflective land surfaces, such as salt flats, a more realistic appearance."[47] The Earth's terrain varies from green to brown. Swirls of clouds congregate over Canada, the North Pole, and over the western coast of South America. The rest of the globe is relatively free from obstructive cloud cover. The royal blue oceans of *Big Blue Marble: Next Generation* are noticeably brighter in intensity than the water masses of the original *Big Blue Marble* and of *The New World* (the 1997 digital version).

Big Blue Marble: Next Generation is the result of significant technological improvements over *The New World* composite photograph from 1997 (which was based on Geostationary-Orbiting Environmental Satellite data), according to NASA Senior Geoscientist William L. Stefanov. Coupled with improved geolocation data, computer hardware, server connectivity, and image processing made it become possible to produce and serve frequently updated, global mosaics at approximately 1 kilometer per pixel or 5 kilometers per pixel resolution.[48] Advances in data processing also improved image clarity, and gave "highly reflective land surfaces, such as salt flats, a more realistic appearance."[49]

In *Big Blue Marble: Next Generation*, technological improvements reduced data-collecting time for images representing every part of the Earth's surface to *one day*, a significant decrease from the *four months* of data collecting needed to make *The New World* in 1997. The imprint of additional time was rendered unnecessary, in favor of producing an image *as quickly as possible*, given newer, "time-saving" technology.[50] This technology steadily evolved until the final whole-Earth composites were made by NASA in 2012. The task of collecting and aggregating data about the surface of the Earth entered the private sector as satellite feeds were released to the public. The introduction in 2011 of Google Earth API (Application Programming Interface) – and, later, Google Earth Engine – obviated NASA's need to dedicate time and resources to producing its own whole-Earth imagery, according to Stefanov. Google Earth evolved into a virtual-travel/entertainment application, while Google Earth Engine markets itself as "[a] planetary-scale platform for Earth science data and analysis."[51] Planet/Google Cloud – the data-collection platform from which Google Earth Engine images are made – claims to harvest seven to ten terabytes of visual data from the Earth's surface each day (equivalent to about 300,000 square kilometers of imagery daily), using about 200 satellites. Access to this data is sold by subscription. Planet's/Google Cloud's customer base exceeds that of only scientists and geologists, and includes the agriculture industry, various governmental bodies (including defense, intelligence, emergency management, energy and infrastructure, forestry, etc.), commercial mapping companies, the insurance industry, commodities trading, and the finance industry.[52] It boasts the capability to create a comprehensive, detailed image of the entire Earth every day, in "near real time" – which, as of June 2018, was available with a ninety-minute delay from "lived" time. (As Planet/Google Cloud launches more satellites to concentrate its image-making efforts, this time is expected to decrease – ideally, to "real time.")

The challenge of decreasing exposure and processing times is hardly one that is unique to the digital age. Rather, it has existed almost as long as the medium of photography itself has.[53] Most survey histories of the medium present a chronological, positivistic story of a series of progressive technological improvements which increased the light-sensitivity of plate and negative chemistry in order to produce images faster, to capture motion, and to reduce or eliminate blur. The medium of photography, whether analog or digital, has a unique culture and history of aspiring to collapse time sensitivity thresholds through advances in photo-chemistry, lens design, camera mechanics, or increased data sensitivity. This trend of "defeating" or "saving" time in photography's exposures, processing, printing, and circulation processes continued into the twentieth century and persists today, as evidenced by the focus on data-collection-time reduction in whole-Earth photography, and the increased capacity to store this data. The medium subscribes wholeheartedly to the positivist spirit and goals of technological modernity and postmodernity. Its aim has been to technologically "annihilate" the time of taking, making, and sharing photographs to enable the near-"real-time," immediate experience of spaces and places worldwide.

In this sense, photography is both a catalyst and symptom of technological modernity and postmodernity. In the book *Liquid Modernity*, sociologist Zygmunt Bauman poses the possibility that modernity itself is a time obsessed with "acceleration and the resulting ability to manipulate time."[54] The acceleration of the production and dissemination of photographic data visualizations – which are highly synthetic "tapestries" of time, combined and distributed digitally – certainly suggests an obsession with "conquering" time with technology. Philosopher and sociologist Jacques Ellul has noted that modernity's near-religious fervor for ever-more-efficient technology, and its increasing emphasis on speed, has had specific social and psychological consequences.[55] *Big Blue Marble: Next Generation* is such an expression of an unrelenting pace of technological advance, one that reflects the fervor for accessing raw data that has been synthesized for quick digestion.

As time in these data visualizations is rendered sculptural and is distended, montaged-together, compressed, and repackaged to form a new idealized, instantaneous version of imaginary time, temporality's very laws, contradictions, and operations are subject to interrogation. Perhaps they even are open to reconfiguration and conquest by technology itself. Dorthe Gert Simonsen has argued that modernity's positivist technological advances, and its increasing valuation of speed, inevitably has led to a convoluting or "annihilation" of our understanding of time:

> [A]cceleration has continuously compressed time and space throughout modernity, reaching a final annihilation of temporal and spatial co-ordinates in the present society of instantaneous electronic communications … Rather than continuously compressing and finally annihilating time and space, speed has produced diverse and interacting, occasionally conflicting, times and spaces.[56]

Composite whole-Earth photographs such as *Big Blue Marble: Next Generation* and the more recent version by the same title made in 2012 (Fig. 5.6) are examples of experimentation with temporality's sensory contingency. In them, we see a fluid, sculptural conception of time as conflictingly non-simultaneous, yet combined in the service of providing a conceptual image that is convincingly realistic as a purveyor of useful data. The time abstractions in digital composite whole-Earth images – and of their analog predecessors – are products of an anxious culture

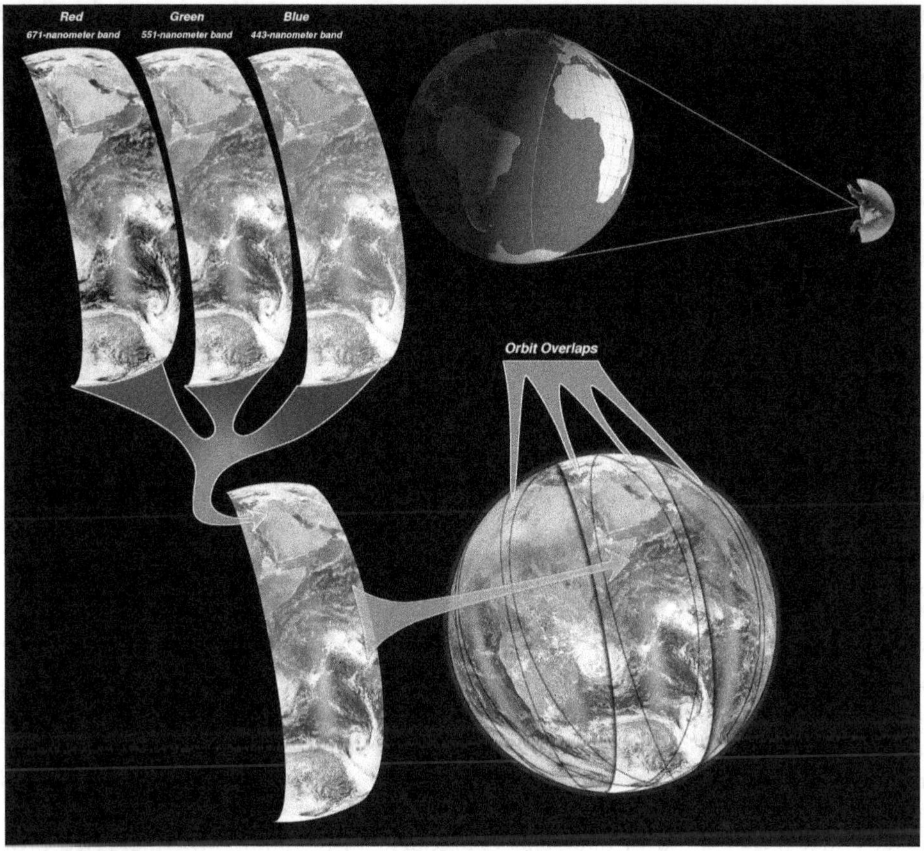

Figure 5.6 National Aeronautics and Space Administration, *Diagram, The Making of* Blue Marble *by Suomi NPP, the Visible Infrared Imaging Radiometer Suite (VIIRS)*, 2012. www.nasa.gov/topics/earth/features/viirs-globe-east.html. Courtesy of the National Aeronautics and Space Administration/Public Domain.

coming to grips with the effects of the quickening pace of life that has been enabled by the same digital tools used in photography to express time's multifarious, fluid experience.

Photography as Medium for Data Visualization: Synthetic, Hypothetical Time

Although NASA's whole-Earth data visualizations present viewers with a seamless digital synthesis of an imaginary, hypothetical, idealized, multi-layered, synthetic, "tapestry" of time, these composite images often are regarded as truthful scientific inscriptions. They are circulated by NASA (and also Google) free of charge online to the public, and also are used as scientific evidence for the study of coastal erosion, flood-zone identification, and the retreat of glaciers and ice caps. These free-of-charge "geovisualizations" allowed scientists to study the impact of human beings upon on the Earth's environment without having to worry about clouds blocking their views of the land during their window of data collection, and without the potential cost of human lives and billions of dollars on launching space missions to gather visual information. (The more recent, for-profit Google Cloud/Planet-generated data visualizations enable access to such data, but for a charge. Still, this charge is more economical for Earth scientists than the startup cost of launching one's own satellites, gathering, and synthesizing the information on their own.)

Whole-Earth images exemplify digital photography's employment for the purpose of creating "data visualizations" – or, visual representations of complex information in straightforward form, with the goal of reinforcing human cognition. Specifically, the piecing together of a tapestry of photographic satellite data from different times and locations, the conversion of this data into mathematical figures, and the translation of this information into a spherical model for further study fulfills the purpose of producing a product based on an *average*, rather than *actual*, appearance. NASA's whole-Earth images are all a function of producing an image of *hypothetical, data visualizations* of the Earth that enjoy the documentary truth value of a medium often associated with representing "the real." The data integrity of whole-Earth visualizations is hardly questioned.

Yet the temporality depicted in a digital composite NASA whole-Earth photograph is a *simulated* time, constructed from tens of thousands of intensely aggregated, digitally synthesized photographs to appear naturalistic. As such, whole-Earth images exemplify what electronic media critic and theorist Kimon Valaskakis has called a state of "informediation," in which an amazingly realistic simulation and synthesis of events and things is possible.[57] Valaskakis also warns that this phenomenon, when not clearly presented as educational, abstract, and hypothetical in its visual presentation of data, has consequences, including "the increased probability of Reality Blur, which may be defined as a progressive blurring of the distinction between truth and fiction, reality and fantasy, actuality and potentiality."[58] This is especially pertinent when the "informediation" in question is produced in the final form of a photograph, a medium with a conflicting, conditional relationship to "the real."

When digital whole-Earth composite photographs appear without explicit captions to explain their basis in abstraction, these images may mislead viewers into mixing fact and fiction, promoting the phenomenon of "Reality Blur." ARC Science Simulations, as mentioned previously, claims that producing "photo-realism" is the intended goal of their (and, by extension, NASA's) whole-Earth photographs. What we see in these digital images, partial products of ARC software and data, are convincingly real abstractions which carry with them an air of scientific truthfulness because they are based on the translation of photographs into data, and back into photographic images again, after vigorous digital processing – and the possibility of human intervention to tweak the appearance of the final results.

By presenting the averaged, hypothetical appearance of the Earth over time, a whole-Earth photograph is an image of the general and the schematic. It is an exemplary image that speaks authoritatively for the visual appearance of the Earth around the vaguely defined timeframe of data collection. The "generalized appearance" offered by a whole-Earth photograph is at odds with what Peter Galassi has argued is photography's unique ability to represent subject-specificity, or "thisness" – following from the logic that the medium is the heir to the "spirit of realism" inherent in visual art since the Renaissance.[59] When used as a medium for data visualizations, photography's rhetoric of realism conveys a statistical "thisness," rather than descriptive, singular, visual "thisness" that corresponds to the subject's actual appearance in a fraction of a moment of exposure time.

Art historians, such as Jonathan Crary, have suggested that modern art was marked by a departure from a sense of the real and indexical, and that a viewing public developed the capacity for appreciating subjectivity:

> From the mid-[nineteenth] century on, an extensive amount of work in science, philosophy, psychology, and art was coming to terms in various ways with the understanding that vision, or any of the senses, could no longer claim an essential objectivity or certainty.[60]

Whole-Earth composite photography's detachment from certainty, specificity, and objectivity, its affinity for the general and the subjective, and its nature (as a product of mathematical and computer science) makes it an example of what Crary argues has fundamentally been at stake in modern vision starting in the nineteenth century. Photography co-evolved with this phenomenon in the nineteenth century, proving itself as a useful tool for pre-digital visualizations of information for the scientific community (which, in England, included the circle of innovators associated with the invention of photography itself). Contact prints of botanical specimens made in the 1830s–1850s by pioneering calotype inventor William Henry Fox Talbot, and by Anna Atkins (who was trained as a botanist and learned the cyanotype process from astronomer Sir John Herschel), may have lacked surface detail and accuracy to the plants' specific hues, but they still were regarded as persuasive documents that testified to the structure and shape of the plants themselves. This was not only a British phenomenon. The microscopic photographs of French physicists Léon Foucault and Hippolyte Fizeau (c. 1845), daguerreotypes of the Moon made with the Great Refractor Telescope by Americans George Bond and John Whipple (c. 1851), and German Wilhelm Röntgen's X-ray images (c. 1895), each represented realities unverifiable and inaccessible by the unaided human eye, but which nonetheless were eagerly accepted by the scientific community as "fact." And Francis Galton's creation of a technique of combining multiple portraits to create a composite image comprised of anywhere from two to several hundred constituent portraits helped to further eugenics (which Galton invented in 1883) by giving mental stereotypes an enduring physical manifestation as a tactile visual image, and empirical "proof" – even though Galton testified that he liberally altered the composition of his original sample set to best recreate the stereotype as it existed in his mind.

Photography as Medium for Analog Visualizations

In the late 1880s in the United States, this practice became a popular technique for group portraiture. Charles O. Lovell's *Composite of Class of '87 Harvard*, 1887 (Fig. 5.7), for instance, is a round-shaped facial portrait of a dark-haired man. Details of his facial features and body form are obscured in a ghostly haze. The

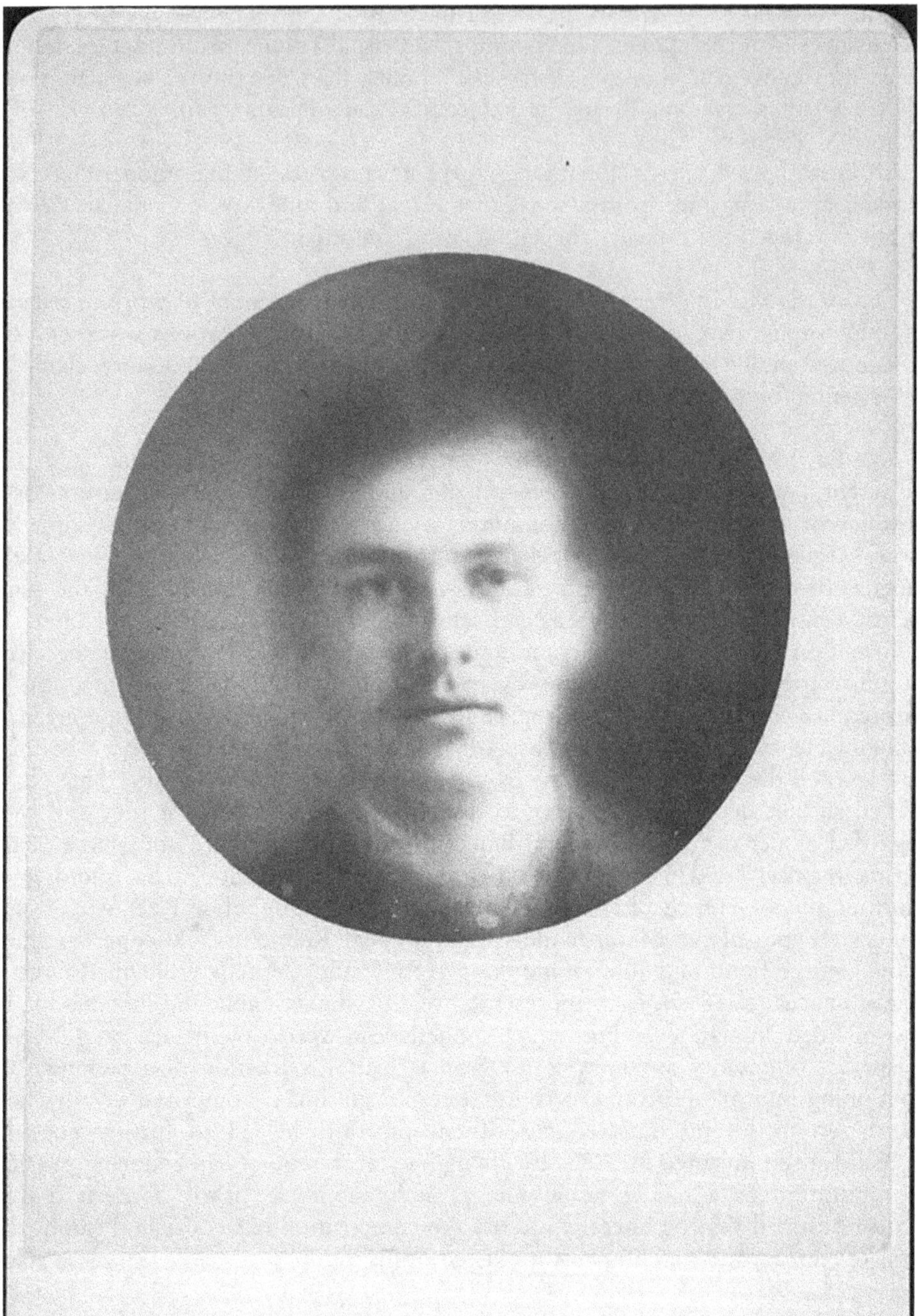

Figure 5.7 Lovell Photo/Charles O. Lovell, *Composite of Class of '87, Harvard University*, 1887. (Harvard University Archives/Public Domain)

image's title indicates that this photograph of what appears to be *one* man actually represents Harvard University's *entire*, all-male graduating class of 1887.[61] Portraits such as this was regarded as an amusing "technical gimmick" that had particular mass appeal in "an age fascinated by the diversity of human appearance and achievement" and in defining "a cultural bedrock of basic human types."[62]

To create this imaginary photograph representing, typologically, Boston's elite, well-bred Brahmin class, Lovell carefully aligned and exposed an unspecified number of portrait negatives on a single sheet of photographic paper.[63] Each individual portrait that comprised the composite image resulted from a fraction-of-a-second exposure, with a specific relationship to an isolated fragment of time from the temporal continuum. However, as these singular images were combined during the printing process, they created a portrait of a human being who never existed, and thus suggested the existence of an instantaneous-exposure-time duration that likewise never existed. Instead, the photograph implies the temporal dimension of an imaginary, abstract "composite time."

Moreover, analog composite portraits such as *Class of '87 Harvard* also embody the temporal implications of the labor-intensive darkroom printing process by which the multiple portrait negatives were aligned and exposed separately on a single sheet of albumen paper. When we view *Class of '87 Harvard*, we see layer over layer of individual fraction-of-a-second exposure times which lose their temporal specificity to produce a singular image that may be comprised of many distinct exposures. But the resulting image paradoxically enjoys no exposure-time-based temporal particularity in and of itself. Instead, the image represents the "average" appearance of Harvard University's well-bred Boston Brahmin brotherhood, a typographic likeness that has resonance in viewers' minds, in the realm of the temporal. They are "real," without being "actual," and make a seemingly unquantifiable phenomenon visually tangible. Analog images such as *Class of '87 Harvard* provide precedents for studying the reception of the unique photographic data visualizations offered by NASA's whole-Earth photographs, and other composite image-making practices.

New England college and university composite portraits such as *Class of '87 Harvard* were prominently displayed as proof of the merits of positive eugenics. Henry Pickering Bowditch published more than a dozen – including the composite portrait of the Smith College Class of 1887 – in the official publication of the Second International Exhibition of Eugenics, alongside images of practicing physicians, mathematicians, members of Bowditch's family, members of the National Academy of Science, Harvard faculty, Wends, and Saxons.[64] Composite portraits gave eugenics the intellectual heft of "science."

The portraits' truth value was premised on what Allan Sekula notes as its status as a condensed version of a full archive, one that "attempts to exist as a potent single image, and the single image attempts to achieve the authority of the archive."[65] Galton's process of making a composite was exacting, rigorous, and even "scientific" in its precision – and, by extension, its perceived "objectivity."[66] Fidelity to the full archive was a prerequisite to a composite photograph's objective integrity. However, Galton notes in his biography that his most successful composite images were comprised almost completely of photographs that "were not greatly unlike, and were of the same size, as judged by measuring the vertical distance between the pupils of the eyes and the parting of the lips."[67] In other words, he altered his archives.

The resulting "objective character" in a composite, Galton suggests, also was a subjective visual encapsulation of the image-maker's preconception of the common physical features shared by a set of individuals. Galton called such typological judgments of sets of people "generic mental images," and conflated the manner in which a photographic plate impresses multiple exposures to the way in which the human sensorium forms impressions of others: "A generic mental image may be considered ... nothing more than a generic portrait stamped on the brain by the cumulative, successive impressions made by its component images."[68] Galton here ascribes the human eye and mind with the capacity to isolate visualized atomized bits of time from the lived continuum, while conflating photography and the temporal process of forming stereotypes.[69]

Historian of data visualization Ingrid Kallick-Wakker argues that knowledge was obtained in the late nineteenth century through a process called "Anschauung," which was inclusive of viewing, inspecting, examining, forming perceptions intuitively, and characterized as "the intuition through pictures formed in the mind's eye from previous visualizations of physical processes in the world of perceptions."[70] Kallick-Wakker further suggests that such subjective images were "often treated as factual reports independent of social and cultural circumstances" by members of various scientific communities.[71] Thus, the denial of societal and cultural influence over the photographer's preconception of "generic mental images" for a finished composite was critical to upholding the image's perceived objectivity, as well as the photographer's. Galton's technique of composite photography enjoyed credibility as a means of translating perceived typologies and stereotypes into tangible pictures that also functioned as statistical averages – or a form of analog biometrics – in accord with then-prevailing "Anschauung"-based processes of knowledge-creation. As Christopher Hookway has argued, the composite photograph "is a schema whose representation in the imagination yields something that can be compared with and/or applied to sensory experience."[72] After all, the "truth" in a scientific data visualization, is relative, as Mimi Cazort suggests:

> Evaluation of the verisimilitude of any scientific image depends on the purpose for which the illustration was made. The camera does not lie, in that the photographic image minimizes the intrusion of the artist's vision. The specimens themselves, however, can "lie."[73]

Data Made Visible: Photographic Composites, Theorized

As a "data portrait" that vacillated between objectivity and subjectivity, this image of social-class celebration was among a few exemplary, mass-accessible, photographic practices that required a conditional understanding of the operations of the photographic document by their viewers as they revealed an early use of the photograph as a means to make information accessible in eras pre-dating the emergence of digital image-making and synthesis.[74] This is to say, since photography's emergence in the early nineteenth century, it has enjoyed a flexible, conditional relationship to "the real," especially when it functioned as a visual purveyor of scientific concepts and visual "averages" – or data visualizations. The digital age merely has brought an expanded set of tools to photographic information-mapping processes that can help generate images with a greater rapidity, flexibility, and ease.

Thus, redefining our understanding of visual objectivity, as a result of the evolution of media, is not a concern unique to the nineteenth century. Historian of media and film Jeffrey Sconce describes the impact of various media – from telegraphy to television – of the modern age as the harbingers of a crisis of "simulation," in which signifiers cease corresponding to objects:

> Where there was once the "real," there is now only the electronic generation and circulation of almost supernatural simulations. Where there was once stable human consciousness, there are now only ghosts of fragmented, decentered, and increasingly schizophrenic subjectivities. Where there was once "depth" and "affect," there is now only "surface." Where there was once "meaning," "history," and a solid realm of "signifieds," there is now only a haunted landscape of vacant and shifting signifiers.[75]

Sconce suggests that the visual communications of various mass media of significantly reordered perceptions of "the real" for a global audience of twentieth-century media in the developed world. Digital whole-Earth photographs, the product of late twentieth-century technology, also operate according to the meaning-production economy Sconce mentions. In these images, the actual Earth ("the signified") is not synonymous with its image ("the signifier"). The Earth, as it appears in whole-Earth digital composites, never existed. But the "signifier" has not shifted, nor is it necessarily vacant. Rather, data – or satellite photographs of hundreds of thousands of images ("signifiers") – is collected and synthesized mathematically to produce a visual photographic tapestry of images of an "averaged" and "simulated" – but not *actual* – appearance of the signified (our planet). (In Peirce's semiotic terms, whole-Earth images operate not as "indexes" of the subject's *actual* appearance as a whole, but as "icons" that share a resemblance with their subject – our planet.) Fredric Jameson similarly notes what he sees as a postmodern play of surface and simulation in visual culture. He argues that four binaries have been collapsed in postmodern culture: "essence and appearance," "latent and manifest," "authenticity and inauthenticity," and "signifier and signified."[76] Digital whole-Earth composites antagonize Jameson's binary of "authenticity and inauthenticity" (by giving convincingly "photo-realistic," and therefore "authentificated," form to the imaginary and abstract, "unauthentic" Earth).

The composite image appears seductively realistic, much like the "supernatural simulations" Sconce sees in previous media. Specifically citing the telegraph and television, he suggests that twentieth-century electronic media changed the public's perception of reality by offering viewers of a subject ("signified") that has only a virtual presence before the viewer. Likewise, the invention of photography enabled a subject to enjoy a two-dimensional, reconstituted, visual presence as a flattened image across multiple spaces and times – without need for the original "signified" (or subject matter) to be physically present, too. The ease and rapidity with which digital photography can be made and shared across multiple nexuses of time and space simply magnifies this effect. Indeed, whole-Earth digital composite photographs are expressions of an age in which the technology-mediated experience of time has been compressed, sped up, made simultaneous, measured, and defined in ever-more-infinitesimal increments. They are the results of complex digital synthesis of visual data, and the layered, elastic representation of time as "hypothetical" or as a "visual average." These *time abstractions* express the anxieties of a culture coming to grips with the ubiquity of accelerating speeds and collapsed time – which are enabled and embodied by the very technology used to make these composite photographs.

But digital composite data visualizations also have inherited dialogs about the medium's conditional relationship to science, to "the real," and to technological modernity and postmodernity that have been a part of the medium's story since its genesis.[77] Digital tools merely have made these temporal conversations more urgent, as near-instantaneous image synthesis and exchange have collapsed time in imaginative new ways.

Notes

1 Some parts of Brand's story first appeared in this essay: Kris Belden-Adams, "Big Blue Marble's Re-Generation: Synthetic Time in NASA's Digital Whole-Earth Photography," *Spectator*, Vol. 28, No. 2 (Fall 2008): 23–30. Special thanks to *Spectator* editor René Bruckner for his helpful edits and feedback.

2 Stewart Brand's role in the development of NASA's whole-Earth photography is detailed in Vicki Goldberg, *The Power of Photography: How Photography Changed Our Lives* (New York: Abbeville Publishing Group, 1991), 53–55. Although Brand's international campaign gained respectable mass support, its influence over NASA is a matter of debate among scholars. Although photography historians such as Goldberg suggest that Brand's mass-marketed drive held sway, Marta Braun disagrees: "NASA wanted high-resolution photographs of proposed lunar landing sites as well as locations of scientific interest … [N]o one had thought of photographing the earth from lunar orbit. Anders' photograph [*Earthrise*], the first one ever made of the earth from deep space, as well as the first to be widely reproduced, was made on impulse." Marta Braun, "Time and Photography," in *Tempus Fugit: Time Flies*, Jan Schall, ed. (Kansas City, MO: Nelson-Atkins Museum of Art, 2000), 141. Information from NASA suggests that photography's scientific credibility as a document held more influence than Brand's campaign, according to National Aeronautics and Space Administration, "The Gateway to Astronaut Photography of the Earth," Earth Sciences and Image Analysis Laboratory, Johnson Space Center, accessed February 7, 2007, available from http://eol.jsc.nasa.gov/, http://eol.jsc.nasa.govFAQ/gen Earth.htm/. However, the medium's reproducibility and its resulting capacity to rouse public enthusiasm for the space program also must have been a consideration in NASA's decision to photograph on its missions.

3 Goldberg, *The Power of Photography*, 54–55.

4 Goldberg, *The Power of Photography*, 54–55.

5 Kevin W. Kelley, ed., *The Home Planet* (Reading, MA: Addison-Wesley, 1988), inside front cover. Hoyle also suggested that seeing our planet diminished by the vast darkness of outer space might humble humankind, while also presenting us with the inspiration for further scientific study of the new frontier of outer space. As Jay Apt, Michael R. Helfert, Justin Wilkinson, and Roger Ressmeyer point out, Socrates may have made this claim even earlier: "Socrates said that if we could rise above the Earth, we would realize 'this is … the true Earth' – and only then would we fully understand the world in which we live." Jay Apt, Michael R. Helfert, Justin Wilkinson, and Roger Ressmeyer, *Orbit: NASA Astronauts Photograph the Earth* (Washington, DC: National Geographic Society, 1996), 11.

6 The *Whole Earth Catalog* became a periodical which was published twice a year from 1968 to 1972, and occasionally thereafter. As of 2018, its most recent publication appeared in 2003.

7 Anders, a good friend of Borman's, ceded authorship to him. It was determined thirty years later that Anders took the photograph. For more about this authorship dispute, see

Robert Zimmerman, "Photo-Finish: A Thirty-Year Controversy Over the Famous Apollo 8 Earthrise Picture Is Finally Resolved," *The Sciences* (1998): 16–18.

8 Apollo 8 On-Board Voice Recorder Transcript, Johnson Space Center Archives, at 75 hours 47 min. This transcript is also published in Robert Poole, *Earthrise: How Man First Saw the Earth* (New Haven: Yale University Press, 2008), 1. For a timeline of image-making of the Earth from points unfixed to the ground, please see Judith Barry, "Map-pings – A Chronology of Remote Sensing," in *Incorporations*, Jonathan Crary and Sanford Kwinter, eds. (New York: Zone, 1992), 570–571.

9 Frank Borman, *Countdown: An Autobiography* (New York: Silver Arrow, 1988), 212.

10 Goldberg, *The Power of Photography*, 56.

11 Goldberg, *The Power of Photography*, 56.

12 Denis Cosgrove, "Contested Global Visions: One-World, Whole-Earth, and the Apollo Space Photographs." *Annals of the Association of American Geographers*, Vol. 84, No. 2 (1994): 283. *Earthrise* inspired widespread public attention to the relatively new environmental movement, which previously only had been popular within the "hippie" counterculture. Dozens of new environmental groups formed, memberships to existing ones soared, and donations to environmental causes reached a new peak. Goldberg, *The Power of Photography*, 57. Before environmental activism became more widespread in America, two pieces of legislation passed, indicating receptiveness on the part of the U.S. government to these issues. In 1963, the Clean Air Act was passed, followed in 1965 by the passage of federal emissions standards for cars: Goldberg, *The Power of Photography*, 53.

13 Braun, "Time and Photography," 140–141.

14 In 1969, however, two American astronauts walked on the Moon, and the event was documented photographically to provide visual proof of the accomplishment for an international audience. Photography provided a means to prove these competitive scientific milestones. For more about the chronology of the space race, and photography's role in it, see: Goldberg, *The Power of Photography*, 52. According to art historian Pamela M. Lee, the Apollo 11 "Moonwalk" photographs sparked a resurgence of a wave of public utopian enthusiasm for technological and scientific positivism, while enabling the perception of photography as a facilitator of such advance: "That picture [of the Moonwalk] has come to stand as among the most profound in the history of technology, offering a view of the 1960s buoyed by the Enlightenment platforms of reason and progress. A utopian vision, perhaps, it is a vision of unflagging optimism, of limitless horizons, and the can-do ethos of American invention"; Pamela M. Lee, *Chronophobia: On Time in the Art of the 1960s* (New York: MIT Press, 2004), 5. Space travel was the subject of great national rivalry between the United States of America and Russia during the Cold War era. The USSR began the "space race" by launching Earth-orbiting spacecraft Sputnik in 1957. They also sent the first human being into outer space for an orbit of the Earth, in 1961. From that point forward, NASA rallied to compete with the Russians. The United States of America led the "space race" in lunar orbitals and landings in the late 1960s and early 1970s.

15 Cosgrove, "Contested Global Visions," 270.

16 Cosgrove, "Contested Global Visions," 278. Denis Cosgrove argues that "representations of the globe and the whole Earth in the twentieth century have drawn upon and constituted a repertoire of sacred and secular, colonial and imperial meanings, and … these representations have played an essentially significant role in the self-representation of the post-war United States and its geo-cultural mission. While the Apollo lunar project signified the achievement of the technocratic goals and universalist rhetoric of Modernism, the project's most enduring legacy is a collection of images whose meanings are contested in post-colonial and postmodernist discourses"; Cosgrove, "Contested Global Visions," 270.

17 NASA estimates that fifty percent of the Earth's surface is covered in clouds in this image: Image Science and Analysis Laboratory, "The Gateway to Astronaut Photography of Earth," Johnson Space Center, Houston, available at http://eol.jsc.nasa.gov/, http://eol.jsc.nasa.gov/scripts/sseop/photo.pl?mission=AS17&roll=148&frame=22727; http://eol.jsc.nasa.gov/FAQ/genEarth.htm; http://svs.gsfc.nasa.gov/vis/a000000/a002680/; http://history.nasa.gov/alsj/a17/images17.html#MagNN (all accessed February 7, 2007).

18 Image Science and Analysis Laboratory, n.p. NASA also created whole-Earth imagery in order to better understand ecosystems on other planets. According to Kamlesh Lulla, Chief Scientist for Earth Observation at Johnson Space Center – who coordinates Earth Observation for Space Station astronauts: "To understand what's happening on Mars, you have to understand what's happening on Earth"; Debbie Nguyen, "Putting the Pieces Together: More Than 100,000 Pictures Have Been Taken From the Space Station," *Space Center Roundup*, Vol. 43, No. 4 (April 2004), 5.

19 This image (titled *Big Blue Marble*, from Apollo 17) is labeled AS17-148–22,727. This naming convention conveys that the photograph is from mission AS17 (or "Apollo 17"), roll 148, frame 22,727.

20 Image Science and Analysis Laboratory, "The Gateway to Astronaut Photography of the Earth."

21 Cosgrove, "Contested Global Visions," 275. Whole-Earth photographs are used by scientists as comparative data to look for evidence of erosion, to track the location and damage of forest fires, to measure the size of hurricanes, and to study the effects of global warming. *Big Blue Marble* also appeared at the start of the movie *An Inconvenient Truth*, former Vice President Al Gore's award-winning documentary on global warming: Davis Guggenheim (director), *An Inconvenient Truth* (Lawrence Bender Productions, 2006).

22 Cosgrove, "Contested Global Visions," 274.

23 Image Science and Analysis Laboratory, "The Gateway to Astronaut Photography of the Earth."

24 Eric Hartwell, *Apollo 17: The Big Marble*. Accessed September 5, 2007, available at www.ehartwell.com/Apollo17/. Upon return, Apollo 17's Preliminary Science Report gave little attention to the Earth photographs. According to Cosgrove: "The popular response was … more enthusiastic. Indeed, the very ubiquity of 22,727 [*Big Blue Marble*], endlessly reproduced on postcards, lapel buttons, flags, calendars, political manifestos, commercial advertisements, and tee-shirts serves to deflect close scrutiny of the image's content. In some cases, its reproducers have unconsciously reversed the image as, for example, when the BBC TV News used 22,727 [*Big Blue Marble*] as the logo for their week-long coverage of the Rio Earth Summit in 1992. All of this suggests that the formal qualities of the image outweighed its cartographic content"; Cosgrove, "Contested Global Visions," 276. It is important to note that the Library of Congress also claims that "Migrant Mother" by Dorothea Lange is the most-requested and -reproduced photograph. For a discussion of the impact of reproduction upon the function of "Migrant Mother," see Robert Hariman and John Louis Lucaites, *No Caption Needed: Iconic Photographs, Public Culture, and Liberal Democracy* (Chicago, IL: University of Chicago Press, 2007), 53–67.

25 Charles Sanders Peirce, *Collected Papers of Charles Sanders Peirce*, Charles Hartstone and Paul Weiss, ed. (Cambridge, MA: Harvard University Press, 1932), 267.

26 Despite a flurry of excitement over the Apollo 17 mission, funding and interest in America's space program have waned since 1972. For more about the decline of interest in America's space program, please see Cosgrove, "Contested Global Visions," 274.

27 Geoffrey Batchen, *Each Wild Idea: Writing, Photography, History* (Cambridge, MA: MIT Press, 2001), 140.

28 For more about the decline of interest in America's space program, see: Cosgrove, "Contested Global Visions," 274.

29 William L. Stefanov, Technical Lead of the Earth Science and Remote Sensing (ESRS) Unit, NASA, interview by the author and email consultation, 2006–2018.

30 NASA has called its digital composite photography of the *Earth Big Blue Marble: Next Generation* since around the year 2000. It converted from using analog photography to document the planet to digital photography in 1991. Stefanov, interview by the author, 2006.

31 ARC Science Simulations, "Face of the Earth," www.arcscience.com/face.htm and www.arcscience.com/gallery.htm (accessed December 9, 2006). According to ARC Science Simulations, "Face of the Earth" data is derived from the "NOAA TIROS satellite AVHRR data and colorized by a proprietary methodology patented by ARC. It has

continued to be enhanced by MODIS and Landsat derivatives"; ARC Science Simulations, "Face of the Earth."

32 ARC Science Simulations, "Face of the Earth."

33 In addition to NASA, these other organizations and companies also used ARC's software and data around the turn of the millennium: the Russian Centre of Cosmonaut Training, the Department of Defense, SPAWAR (Space and Naval Warfare Systems, USA), UNAVCO (University NAVSTAR Consortium, USA), Raytheon, and Lockheed Martin. ARC also sells programs for making images of Mars, Jupiter, the Moon, and for mapping the surface appearance of the Earth up to 600 million years ago: ARC Science Simulations. Anyone with about $8,200 to spare in 2018 – and a lot of time to learn the intricacies of the complicated computer program – can purchase software to generate a whole-Earth image through a company such as ARC Science Simulations. The Museum of Natural History and the Smithsonian Institution, for instance, have featured ARC images in their exhibitions.

34 This proves most difficult when photographing the polar ice caps, which cannot easily be distinguished from above the cloud covering. ARC Science Simulations.

35 The first data satellites were dispatched in 1972.

36 ARC Science Simulations.

37 All photographs are labeled by date, geographic coordinates, sun elevation angle, sun azimuth, type and brand of sensor, and scene identification number. Additional fields of information may be added to digital images, as deemed necessary by their processors. Thomas Eugene Avery explains the nature of the conversion of "In physical form, a digital image is a two-dimensional array of small areas called pixels, or pels (contractions of *picture elements*), that correspond spatially to relatively small terrain areas called ground resolution cells. The horizontal rows of pixels are called lines, and the vertical columns of pixels are termed samples … Because of this ordering, the origin of a grid referencing system for a digital image is always the upper left pixel; its coordinates are line 1, sample 1"; Thomas Eugene Avery, *Fundamentals of Remote Sensing and Airphoto Interpretation* (New York: Macmillan Publishing Company, 1992), 406. Data may be "corrected" for several reasons. Each satellite's data are slightly and uniquely skewed with respect to the angle of their view of the Earth. Skew is corrected mathematically – according to Avery, "Each scene is deskewed by an algorithm that shifts scan lines by a calculated number of pixels – a number dependent on the estimated latitude for the scan line being processed with respect to the starting line of the image"; Avery, *Fundamentals of Remote Sensing*, 414. Satellite data also have "nonsystematic distortions," including the satellite's roll, pitch, and yaw. The resulting image is compared to a map and corrected; Avery, *Fundamentals of Remote Sensing*, 414. Blurring is also solved by using three correctional algorithms to "average" the nearby data and add it in place of the missing details; Avery, *Fundamentals of Remote Sensing*, 414–416. Sometimes, satellite images have periodic six-line stripes added accidentally by the scanner, specks of scanning "noise," or lines of data are missing. These imperfections are fixed by using a complicated mathematical formula (the digital number of input pixel at line J, sample S multiplied by the product of the standard deviation of the entire scene divided by the standard deviation of individual detector. That number is added to the mean DA value of the entire scene minus the mean DA value of the individual detector – after the mean DA value of the individual detector is multiplied by the standard deviation of the entire scene divided by the standard deviation of the individual detector)"; Avery, *Fundamentals of Remote Sensing*, 417.

38 National Aeronautics and Space Administration, "Visible Earth/Big Blue Marble" NASA Goddard Space Flight Center, Earth Observatory: https://visibleearth.nasa.gov/view.php?id=57723 (accessed June 7, 2018).

39 For instance, the Museum of Natural History, one of ARC Science Simulations' clients, wanted to display a globe revealing land masses – especially the home country of most of its visitors, the United States of America. To create such an image, a technician overrides the "probable appearance" of the composite image to create a globe with lighter-than-mathematically determined cloud covering to reveal more of the desired land mass; ARC Science Simulations, "Face of the Earth."

40 To read more about the way that "decision trees" operate to simulate cloud cover, see Smadar Shiffman, "Cloud Detection from Satellite Imagery: A Comparison of Expert-Generated and Automatically-Generated Decision Trees," *The Intelligence Report* (2005).; Various Articles/Authors, *Science.gov*, www.science.gov/topicpages/a/automated+cloud+detection.html (accessed April 19, 2018).
41 Stefanov, interview by the author, 2006–2018.
42 Reto Stöckli, "The Big Blue Marble Next Generation: A True Color Earth Dataset Including Seasonal Dynamics from Modis," NASA Goddard Space Flight Center, Earth Observatory, http://earthobservatory.nasa.gov/Newsroom/BlueMarble/bmng.pdf. (accessed March 13, 2007).
43 Reto Stöckli, "The Big Blue Marble Next Generation," n.p.
44 ARC's data is derived from the "NOAA TIROS satellite AVHRR data and colorized by a proprietary methodology patented by ARC. It has continued to be enhanced by MODIS and Landsat derivatives"; Stöckli, interview by the author, 2006–2018.
45 Mirzoeff, Nicholas. *The Visual Culture Reader*, second edition (New York: Routledge, 2009), 5.
46 Stefanov, interview by the author and email consultation, 2006.
47 Stöckli, "Blue Marble: Next Generation," n.p. Computer technicians are working to correct what they see as problems with *Big Blue Marble: Next Generation* images. Stöckli argues that "the creator of the *Blue Marble* [software] uses a uniform blue color for deep ocean regions, and this value has not been completely blended with observations of shallow water in coastal areas." Stöckli also claims that the images also fail to distinguish between snow covering and clouds. He also points out that, during periods of heavy rainfall in tropical regions, it is often difficult to get data about the Earth's surface appearance because of persistent cloud cover; Stöckli, "Blue Marble: Next Generation."
48 Stefanov, interview by the author, 2006–2018.
49 Reto Stöckli, "Blue Marble: Next Generation," NASA, http://earthobservatory.nasa.gov/Newsroom/BlueMarble/ (accessed March 13, 2007).
50 This phenomenon has limitations, computer-technology historian Alex Steffen suggests: "The way we code software also has profound limitations"; Alex Steffen, "What Happens When Technology Zooms Off the Charts?: Singularity and Its Meanings," *Whole Earth Catalog* (Spring 2003): 9. Steffen speculates, based on the state of data storage in 2003, that data processing speeds will reach a peak of efficiency in the years 2010–2020.
51 N. Gorelick, M. Hancher, M. Dixon, S. Ilyushchenko, D. Thau, and R. Moore, "Google Earth Engine: Planetary-Scale Geospatial Analysis for Everyone," *Remote Sensing of Environment* (2017). Available at https://earthengine.google.com/.
52 Planet/Google Cloud, "Frequently Asked Questions," *Planet.com*, available at www.planet.com/faqs/s. (accessed June 7, 2018). Planet was founded in 2010 by three NASA scientists. Planet/Google Cloud offers a gallery of images that presumably would be of interest to customers. In it, the public can find images of floods, fires, a Tesla factory, Apple's campus, mines, cities in India, the Disneyland parking lot, cities worldwide, dams, volcanoes, space centers in India, ports worldwide, oil-processing facilities in Saudi Arabia, a Chinese solar farm, rivers, illegal mining in Peru, and sugarcane deforestation in Bolivia.
53 Although the reduction of exposure times was a particular preoccupation in the nineteenth century, a collapse of processing time analogous to the impact of digital processing was made possible in 1947 with the introduction of the Polaroid Land Camera.
54 Dorthe Gert Simonsen, "Accelerating Modernity: Time-Space Compression in the Wake of the Aeroplane." *Journal of Transport History*, Vol. 26, No. 2 (September 2005): 98.
55 Jacques Ellul, *The Technological Society*. John Wilkinson, trans. (New York: Knopf, 1964).
56 Simonsen, "Accelerating Modernity," 99. Simonsen compares the obsession with speed during the Information Age with the similar drive for speed at the time of the invention of the airplane: "Phrases such as 'conquering,' 'shrinking,' and 'annihilation' of space and time have been used in popular discourse to capture the experience of more than a century of accelerating technology. From this perspective 'time-space compression,' which

has been among the tropes most often repeated in the rhetoric of globalization and the Information Age, begins to sound like an echo of the former period"; Simonsen, "Accelerating Modernity," 102.

57 Kimon Valaskakis, "The Concept of Informediation: A Framework for a Structural Interpretation of the Information Revolution," in *Information Technology: Impact on the Way of Life*, ed. Liam Bannon, From the EEC Conference on the Information Society, Dublin, Ireland, November 18–20, 1981 (Dublin: Tycooly International Publishing Ltd., 1982), 33.
58 Valaskakis, "The Concept of Informediation," 31.
59 Peter Galassi, *Before Photography: Painting and the Invention of Photography* (New York: Museum of Modern Art, 1981), 19.
60 Jonathan Crary, "Unbinding Vision," *October* 68 (1994): 21.
61 Women were educated by Harvard University faculty through the "Harvard Annex." John L. Lovell also created a separate composite photograph of the women who studied at the Annex in 1887. The Annex was renamed "Radcliffe College" two years later, and was chartered as such in 1894.
62 Keith Davis, *An American Century of Photography: From Dry-Plate to Digital*, second edition (New York: Harry N. Abrams, 1999), 24.
63 Davis, *An American Century of Photography*, 24. An informal count of the number of undergraduates who received their Bachelor of Arts degrees in 1887 totals sixty-five. It is not discernible whether, for instance, graduate students or other majors were included in John L. Lovell's multiple-exposed composite photograph. For more about this image's connections to Eugenics movement in the United States, please see Kris Belden-Adams, "Harvard's Composite 'Class' Pictures," *Photographies*, Vol. 8, No. 2 (2015): 125–136, available at https://doi.org/10.1080/17540763.2015.1054300.
64 Mark Haller has argued that the New England class of semi-aristocrats saw their genesis in the Anglo-Saxon lineage: "The Anglo-Saxons, it was held, were an exceptionally restless, industrious, and creative people. Their love of liberty and respect for law originated among Teutonic tribes in the forests of Germany and from there spread with the Anglo-Saxons to England"; Mark H. Haller, *Eugenics: Hereditarian Attitudes in American Thought* (New Brunswick, NJ: Rutgers University Press, 1963), 51.
65 Allan Sekula, "The Body and the Archive," *October*, Vol. 39 (Winter 1986): 54.
66 Composite photography "had wide currency in scientific circles at the time at which Peirce was writing"; Christopher Hookway, "'... a Sort of Composite Photograph': Pragmatism, Ideas, and Schematism," *Transactions of the Charles S. Peirce Society*, 38, Nos. 1–2 (Winter/Spring 2002): 29. Composite photography enjoyed scientific regard as Peirce was conceptualizing his theories describing the functions of signs. Hookway therefore argues that composite portraits likely influenced Peirce's writing on photography.
67 Francis Galton, *Memories of My Life* (London: Methuen, 1908), 260.
68 Francis Galton, *Generic Images* (London: William Clowes and Sons, 1879), 166.
69 More recent studies of brain function argue otherwise: "We don't create representations of visual stimuli, store them in a short-term memory buffer, and then transfer the representation into a long-term memory device. We don't retrieve information or images or words from memory registers. Computers do all of these things, but organisms do not"; Robert Epstein, "'The Empty Brain' Your Brain Does Not Process Information, Retrieve Knowledge or Store Memories," *Aeon*, May 18, 2016. n.p., available at https://aeon.co/essays/your-brain-does-not-process-information-and-it-is-not-a-computer (accessed June 2, 2016).
70 Ingrid Kallick-Wakker, "Science Icons: The Visualization of Scientific Truths," *Leonardo*, Vol. 27, No. 4 (1994): 310.
71 Kallick-Wakker, "Science Icons," 310.
72 Hookway, "'... a Sort of Composite Photograph,'"40.
73 Mimi Cazort, "Photography's Illustrative Ancestors: The Printed Image," in *Beauty of Another Order: Photography in Science*. Ann Thomas, ed. (New Haven: Yale University Press, 1997), 25.
74 Josh Ellenbogen, "Photography and the Imperceptible: Bertillon, Galton, Marey," Ph.D. diss., University of Chicago, 2005; Josh Ellenbogen, "Authority, Objectivity, Evidence:

Scientific Photography in Victorian Britain," *Studies in the History and Philosophy of Science*, Vol. 39 (Spring 2008); Josh Ellenbogen, "Camera and Mind," *Representations*, Vol. 101 (Winter 2008): 86–115.

75 Jeffrey Sconce, *Haunted Media: Electronic Presence From Telegraphy to Television* (Durham, NC: Duke University Press, 2000), 170–171.

76 He also argues that these binary collapses signal the end of "High Modernism" and "High Capitalism": Fredric Jameson, "The End of Temporality," *Critical Inquiry*, Vol. 29 (2003); Frederic Jameson, *Postmodernism, or, the Cultural Logic of Late Capitalism* (Durham, NC: Duke University Press, 1991).

77 Vicki Goldberg has suggested that "The photographic lie succeeds because photographs appear to be so truthful. The fake disrupts two dearly held expectations: that photographs report what was actually there, and that seeing is believing – for photography amounts to a surrogate for personal observation"; Goldberg, *The Power of Photography*, 89. Photography's rhetoric of "the real" indeed makes it a seductive medium with which to produce science simulations.

Bibliography

Ades, Dawn. "Art and Time in the Twentieth Century." In *The Story of* Time, exhibition catalogue. London: Merrell Holberton, 1999. 202–211.

Apt, Jay, et al. *Orbit: NASA Astronauts Photograph the Earth*. Washington, DC: National Geographic Society, 1996.

ARC Science Simulations, "Face of the Earth," www.arcscience.com/face.htm and www.arcscience.com/gallery.htm (accessed December 9, 2006).

Avery, Thomas Eugene. *Fundamentals of Remote Sensing and Airphoto Interpretation*. New York: Macmillan Publishing, 1992.

Bardis, Antonia. "Digital Photography and the Question of Realism." *Journal of Visual Art Practice*, Vol. 3, No. 3 (2004): 209–218.

Barry, Judith. "Mappings – A Chronology of Remote Sensing." In *Incorporations*, Jonathan Crary and Sanford Kwinter, eds. New York: Zone Books, 1992. 570–571.

Baudrillard, Jean. "The Implosion of Meaning in the Media and the Implosion of the Social in the Masses." In *The Myths of Information*, Kathleen Woodward, ed. Madison, WI: Coda Press, 1980.

Baudrillard, Jean. *Simulations*. New York: Semiotext(e), 1983.

Bauman, Zygmunt. *Liquid Modernity*. Cambridge, MA: Polity Press, 2000.

Beardsworth, Richard. "Thinking Technicity." In *Deconstruction: A Reader*, Martin McQuillan, ed. Edinburgh: Edinburgh University Press, 2000. 235–243.

Belden-Adams, Kris. "Big Blue Marble's Re-Generation: Synthetic Time in NASA's Digital Whole-Earth Photography" *Spectator*, Vol. 28, No. 2, Special Issue: "The Instant" (Fall 2008): 23–30.

Belden-Adams, Kris. "Harvard's Composite 'Class' Pictures," *Photographies*, Vol. 8, No. 2 (2015): 125–136.

Belting, Hans. "Image, Culture, History, Media, Visualization, Art." *Critical Inquiry*, Vol. 31, No. 2 (2005): 302–319.

Betsky, Aaron, ed. *010101: Art in Technological Times*. San Francisco, CA: San Francisco Museum of Modern Art, 2001.

Bird, J. Brian. "Space Photography and Its Geographical Applications." *Geographical Review*, Vol. 54, No. 4 (1964): 463–486.

Borchert, Donald M., and Stewart, David, eds. *Being Human in a Technological Age*. Athens: Ohio University Press, 1979.

Frank Borman, *Countdown: An Autobiography*. New York: Silver Arrow, 1988.

Brand, Stewart, ed. *Whole Earth Software Catalog*. Garden City, NY: Quantum Press/Doubleday, 1984.

Castells, Manuel. *The Rise of the Network Society*. Malden, MA: Blackwell Publishers, 1996.

Cosgrove, Denis. "Contested Global Visions: One-World, Whole-Earth, and the Apollo Space Photographs." *Annals of the Association of American Geographers*, Vol. 84, No. 2 (1994): 270–294.

Debord, Guy. *The Society of the Spectacle*. Donald Nicholson-Smith, trans. New York: Zone Books, 1994.

Diebold, John. *Man and the Computer: Technology as an Agent of Social Change*. New York: Frederick A. Praeger, 1969.

Druckey, Timothy, ed. *Electronic Culture: Technology and Visual Representation*. New York: Aperture Foundation, 1996.

Ellenbogen, Josh. "Authority, Objectivity, Evidence: Scientific Photography in Victorian Britain," *Studies in the History and Philosophy of Science*, Vol. 39 (Spring 2008): 171–175.

Ellenbogen, Josh. "Photography and the Imperceptible: Bertillon, Galton, Marey," PhD diss, University of Chicago Press, 2005.

Ellenbogen, Josh. "Camera and Mind," *Representations*, Vol. 101 (Winter 2008): 86–115.

Ellul, Jacques. *The Technological Society*. John Wilkinson, trans. New York: Knopf, 1964.

Ferris, Timothy. *Space Shots*. New York: Pantheon Books, 1984.

Friday, Jonathan. "Photography and the Representation of Vision." *Journal of Aesthetics and Art Criticism*, Vol. 59, No. 4 (2001): 351–362.

Gallery, Henry Art. *Radical, Rational, Space, Time*. Seattle, MA: Henry Gallery Association, 1983.

Gehlen, Arnold. *Man in the Age of Technology*. New York: Columbia University Press, 1980.

Gere, Charlie. *Art, Time and Technology*. Oxford and New York: Berg, 2006.

Giloth, Copper and Pollock-Williams, Lynn. "A Selected Chronology of Computer Art: Exhibitions, Publications, Technology." *Art Journal*, Vol. 49, No. 3 (1990): 283–297.

Gips, Terry. "Computers and Art: Issues of Content." *Art Journal*, Vol. 49, No. 3 (1990): 228–232.

Grosz, Elizabeth, ed. *Becomings: Explorations in Time, Memory, and Futures*. Ithaca, NY, and London: Cornell University Press, 1999.

Haller, Mark H. *Eugenics: Hereditarian Attitudes in American Thought*. New Brunswick, NJ: Rutgers University Press, 1963.

Hookway, Christopher. "'. . .a Sort of Composite Photograph': Pragmatism, Ideas, and Schematism," *Transactions of the Charles S. Peirce Society*, Vol. 38, No. 1–2 (Winter/Spring 2002): 29.

Husserl, Edmund. *The Phenomenology of Internal Time Consciousness*. Martin Heidegger, ed. Bloomington, IN and London: Indiana University Press, 1966.

Husserl, Edmund. *The Crisis of the European Sciences and Transcendental Phenomenology: An Introduction to Phenomenological Philosophy*. David Carr, trans. Evanston, IL: Northwestern University Press, 1970.

Huyssen, Andreas. *After the Great Divide: Modernism, Mass Culture, Postmodernism*. Bloomington, IN, and Indianapolis: Indiana University Press, 1987.

Huyssen, Andreas. *Twilight Memories: Marking Time in a Culture of Amnesia*. London and New York: Routledge, 1994.

Huyssen, Andreas. "Present Pasts: Media, Politics, Amnesia." *Public Culture*, Vol. 12, No. 1 (2000): 21–38.

Ihde, Don. *Philosophy of Technology: An Introduction*. New York: Paragon House, 1993.

Ippolito, Jon. "Trusting Aesthetics to Prosthetics." *Art Journal*, Vol. 56, No. 3 (1997): 68–74.

Jalobeanu, Andre, Kuehnel, Frank O., and Stutz, John C. "Modeling Images of Natural 3-D Surfaces: Overview and Potential Applications." *The Intelligence Report* (2004). Washington, DC. Proceedings from the 2004 Conference on Computer Vision and Pattern Recognition Workshop, Institute of Electrical and Electronics Engineers. doi:10.1109/CVPR.2004.398

Jameson, Fredric. *Postmodernism, or, the Cultural Logic of Late Capitalism*. Durham, NC: Duke University Press, 1991.

Jameson, Fredric. "The End of Temporality." *Critical Inquiry*, Vol. 29 (2003): 695–718.

Jammer, Max. *Concepts of Space: The History of Theories of Space in Physics*. Cambridge, MA: Harvard University Press, 1969.

Johnson, Geraldine A. *Sculpture and Photography: Envisioning the Third Dimension*. Cambridge, UK, and New York: Cambridge University Press, 1998.

Johnston, Adrian. *Time Driven: Metapsychology and the Splitting of the Drive*. Evanston, IL: Northwestern University Press, 2005.

Kallick-Wakker, Ingrid. "Science Icons: The Visualization of Scientific Truths," *Leonardo*, Vol. 27, No. 4 (1994): 310.

Knight, Diana. *Barthes and Utopia: Space, Travel, Writing*. Oxford: Clarendon Press, 1997.

Kuipers, Dean, and Aitken, Doug. *I am a Bullet: Scenes From an Accelerating Culture*. New York: Crown Publishers. 2000.

Laboratory, Image Science and Analysis. *The Gateway to Astronaut Photography of Earth*. Houston, TX: Johnson Space Center. Available from http://eol.jsc.nasa.gov/, http://eol.jsc.nasa.gov/scripts/sseop/photo.pl?mission=AS17&roll=148&frame=22727, http://eol.jsc.nasa.gov/FAQ/genEarth.htm and http://svs.gsfc.nasa.gov/vis/a000000/a002680/, and http://history.nasa.gov/alsj/a17/images17.html#MagNN (accessed February 7, 2007).

Laidler, Keith James. *To Light Such a Candle: Chapters in the History of Science and Technology*. Oxford and New York: Oxford University Press, 1998.

Lee, Pamela M. *Chronophobia: On Time and Art of the 1960s*. Cambridge, MA, and London: MIT Press, 2004.

Lee, Pamela M. "Some Kinds of Duration: The Temporality of Drawing as a Process Art." In *Afterimage: Drawing through Process*, Cornelia H. Butler, ed. Cambridge, MA: MIT Press, 1999.

Leighten, Patricia. "Critical Attitudes Toward Overtly Manipulated Photography in the 20th Century." *Art Journal*, Vol. 37, No. 2 (1977–1978): 133–138.

Levinson, Paul. *The Soft Edge: A Natural History and Future of the Information Revolution*. London and New York: Routledge, 1997.

Lienhard, John. *Inventing Modern: Growing Up With X-rays, Skyscrapers, and Tailfins*. New York: Oxford University Press, 2003.

Lovejoy, Margot. "Art, Technology and Postmodernism: Paradigms, Parallels and Paradoxes." *Art Journal*, Vol. 49, No. 3 (1990): 257–265.

Lunenfeld, Peter, ed. *The Digital Dialectic: New Essays on New Media*. Cambridge, MA: MIT Press, 1999.

Lyotard, Jean-François. *The Postmodern Condition*. Minneapolis: University of Minnesota Press, 1984.

Mercedes, Dawn. "Digital Ethics: Computers, Photographs and the Manipulation of Pixels." *Art Education*, Vol. 49, No. 3 (1996): 44–50.

Merloo, Joost. *Along the Fourth Dimension; Man's Sense of Time and History*. New York: John Day Co., 1970.

Milne, E. A. *Some Points in the Philosophy of Physics: Time, Evolution and Creation*. Washington, DC: Smithsonian Institution Report, 1933.

Miquel, Paul-Antoine. "Evolution of Consciousness and Evolution of Life." *MLN*, Vol. 120, No. 5 (2005): 1156–1167.

Mirzoeff, Nicholas. *The Visual Culture Reader*, second edition. New York: Routledge, 2009.

Morris, Edward, ed., *Constable's Clouds: Paintings and Cloud Studies by John Constable*. Edinburgh and Liverpool: National Galleries of Scotland and the National Museums and Galleries on Merseyside, 2000.

Morris, Robin D., Smelyanskiy, Vadim N., and Cheesman, Peter C. "Matching Images to Models - Camera Calibration for 3-D Surface Reconstruction." *The Intelligence Report*

(2001). Washington, DC; Proceedings from the 2004 Conference on Computer Vision and Pattern Recognition Workshop, Institute of Electrical and Electronics Engineers. doi:10.1109/CVPR.2004.398

Moser, Mary Anne. *Immersed in Technology: Art and Virtual Environments*. Cambridge, MA: MIT Press, 1996.

National Aeronautics and Space Administration. "Visible Earth: A Catalog of NASA Images and Animations of Our Home Planet. Johnson Space Center. Available from http://visible earth.nasa.gov/view_rec.php?id=58 (accessed February 13, 2007).

National Aeronautics and Space Administration. *Visible Earth: A Catalog of NASA Images and Animations of Our Home Planet*. Houston, TX: Johnson Space Center, available from http://visibleearth.nasa.gov/view_set.php?sensorNAME=MODIS (accessed March 13, 2007).

Nguyen, Debbie. "Putting the Pieces Together: More Than 100,000 Pictures Have Been Taken from the Space Station." *Space Center Roundup*, Vol. 43, No. 4 (April 2004).

Olkowski, Dorothea E. "Time Lost, Instantaneity and the Image." *Parallax: A Journal of Metadiscursive Theory and Cultural Practices*, Vol. 9, No. 1 (2003): 28–38.

Ong, W. *Oralty and Literacy: The Technologising of the Word*. London: Methuen, 1982.

Osborne, Peter. "Photography in an Expanding Field: Distributive Unity and Dominant Form." In *Where is the Photograph?* D. Green, ed. Brighton: Photoforum and Photoworks, 2003. 47–62.

Ott, John. *My Ivory Cellar: The Story of Time-Lapse Photography*. Chicago, IL: Twentieth Century Press, 1958.

Perchick, Max. "Extending the Frontiers of Photographic Vision." *PSA Journal*, Vol. 62, No. 11 (1996): 26.

Peters, John Durham. *Speaking Into the Air: A History of the Idea of Communication*. Chicago: University of Chicago Press, 1999.

Poole, Robert. *Earthrise: How Man First Saw the Earth*. New Haven, CT: Yale University Press, 2008.

Popper, Karl R. *The Open Universe: An Argument for Indeterminism*. Totowa, NJ: Rowman & Littlefield, 1982.

Ritchin, Fred. "Photojournalism in the Age of Computers." In *The Critical Image: Essays on Contemporary photography*, Carol Squiers, ed. Seattle, MA: Bay Press, 1990. 28–37.

Robins, Kevin. "The Virtual Unconscious in Photography." In *Electronic Culture*, Timothy Druckey, ed. New York: Aperture Foundation, 1996. 154–163.

Shiffman, Smadar. "Cloud Detection From Satellite Imagery: A Comparison of Expert-Generated and Automatically Generated Decision Trees." *The Intelligence Report* (2005). Washington, DC: Institute of Electrical and Electronics Engineers.

Simonsen, Dorthe Gert. "Accelerating Modernity: Time-Space Compression in the Wake of the Aeroplane." *Journal of Transport History*, Vol. 26, No. 2 (September 2005): 98–117.

Star, Susan Leigh. *The Cultures of Computing*. Cambridge: Blackwell Publishers, 1995.

Stefanov, William L. Technical Lead of the Earth Science and Remote Sensing (ESRS) Unit, NASAnterview by the author and email consultation, 2006–2018.

Steffen, Alex. "What Happens When Technology Zooms Off the Charts? Singularity and Its Meanings," *Whole Earth Catalog*, Spring 2003.

Stiegler, Bernard. *Technics and Time, 1: The Fault of Epimetheus*. Stanford, CA: Stanford University Press, 1998.

Stöckli, Reto. "The Big Blue Marble Next Generation – A True Color Earth Dataset Including Seasonal Dynamics from MODIS." (2005). National Aeronautics and Space Administraion, Goddard Space Flight Center, Earth Observatory, available from http://earthobservatory.nasa.gov/Newsroom/BlueMarble/bmng.pdf (accessed March 13, 2007).

Stöckli, Reto. "Blue Marble: Next Generation." National Aeronautics and Space Administration, Goddard Space Flight Center, Earth Observatory, available from http://earthobserva tory.nasa.gov/Newsroom/BlueMarble/ (accessed March 13, 2007).

Stone, Rosanne Allucquere. *The War of Desire and Technology at the Close of the Mechanical Age*. Cambridge, MA: MIT Press, 1995.
Sturgis, Alexander. *Telling Time*. London: National Gallery Company, 2000.
Sutton, Damien. *Photography, Cinema, Memory: The Crystal Image of Time*. Minneapolis: University of Minnesota Press, 2009.
Sutton, Damien. "Delay. Notes on Photography as Non-Representational Thinking." In *Time Expanded*. Madrid: La Fábrica/PHotoEspaña, 2010. 145–151.
Talbott, Stephen L. *The Future Does Not Compute: Transcending the Machines in Our Midst*. Sebsatopol: O'Reilly and Associates, 1995.
Thomas, Ann. *Beauty of Another Order: Photography in Science*. New Haven, CT: Yale University Press, 1997.
Thomas, Ann. "Capturing Light: Photographing the Universe." In *Beauty of Another Order: Photography in Science*. New Haven, CT: Yale University Press, 1997. 186–219.
Thomas, Ann. "The Search for Pattern." In *Beauty of Another Order: Photography in Science*. New Haven, CT: Yale University Press, 1997. 76–119.
Turkle, Sherry. *Life on the Screen: Identity in the Age of the Internet*. New York: Simon & Schuster, 1995.
Valaskakis, Kimon. "The Concept of Informediation: A Framework for a Structural Interpretation of the Information Revolution." In *Information Technology: Impact on the Way of Life*, Liam Bannon, ed. Dublin: Tycooly International Publishing, 1982.
Winner, L. *Autonomous Technology: Technics-Out-of-Control as a Theme in Political Thought*. Cambridge, MA: MIT Press, 1977.
Zagorin, Perez. "Historiography and Postmodernism: Reconsiderations." *History & Theory*, Vol. 29, No. 3 (1990).
Ziegler, Oswald, ed. *Australian Photography*. Sydney: Ziegler Gotham Publications, 1947.

6 Conclusions

Many commentators have remarked that photography has a unique relationship to time. Many scholars have been content to discuss the medium's expression of temporality in broad, generalized terms, or within the context of the normative "fraction-of-a-moment" conception of photographic time. However, through a detailed and specific examination of the temporal dimensions of a variety of ubiquitous, mass-accessible, yet-unstudied, photographic practices, this book has argued that the medium possesses a multifarious range of ways to express the experience of time.

Chapter 2 featured a look at Edgerton's stroboscopic photographs, which are regarded as archetypes of our conceptions of the medium's conventional, fraction-of-a-second relationship to exposure time. These images also represent a technological milestone and an apex in the medium's technological journey to master the temporal fragment. Moreover, a detailed examination of the various frames of Edgerton's *Death of a Light Bulb/.30 Caliber Bullet* revealed this series to express much more than scientific/technical advance, or an "unreal realism" involving everyday objects. Edgerton's *Death of a Light Bulb* instead revealed that even the most "instantaneous" of photographs, with exposure durations of a millionth of a second, possess a wealth of temporal associations far beyond only that of exposure duration, or a monolithic *that-which-was*.

A close examination of Edgerton's *Death of a Light Bulb* argued that the drive to position photographs within a temporal order and fit into a narrative may supersede the veracity of the photographs themselves. Viewers are quick to temporally contextualize the photographs as illustrations of the linear event Edgerton suggests – the death of *a* light bulb – rather than see the series as a representation of the "deaths" of four different bulbs. The urge to place photographs within a meaningful temporal framework, in order to make sense of time, curiously overpowers the documentary authority of these photographs and poses a challenge to the perceived "truth" value of the medium. This also prompts us to consider the strangeness of photography's images of time, in relationship to our ongoing, unceasing "lived time" experience.

Edgerton's photographs, this project has argued, also reveal the layered temporal nature of *narrative time* – a concept that received an even more extensive examination in Chapter 3, through a case study of a commonplace, exemplary, fraction-of-a-second-duration-exposure, American family portrait, *Leighton Family* (1899). The ubiquitous photographic practice of studio portraiture allows for the investigation of the strangeness of the isolated, *that-has-been* view of full-field, fraction-of-a-second exposure, with respect to the ongoing temporal continuum. Even a photograph such as the *Leighton Family* – which complies with conventional expectations of "photographic time" – forces a radical collision of the past and the lived-time "present," revealing the medium's unique capacity to express the complex experience of time.

Additional temporal complexities emerge as multiple viewers of the *Leighton Family* readily impose varying, mutable, fractured, creative, and interpretive *narrative times* upon the portrait, in order to create a personal and subjective sense of order for the image. These vernacular histories are partial, mutable, innately personal, and rife with omissions and the addition of absent items and/or events. As Olin has pointed out, Barthes made similar "mistakes" in his descriptions of several images in *Camera Lucida.*[1] These errors, whether intentional or not, illuminate discussions of the creative, amalgamated narratives that lend photographs a new, personalized relationship to time. Barthes suggests that these narratives are worthy of study in our photography histories. This book provides further expansion of Barthes's ideas. Moreover, it suggests the merit of continued inquiry into the phenomenon of *narrative time*, not only as a manifestation of temporality, but also as a means of more thoroughly exploring the function and persisting personal relevance of vernacular portraiture practices.

Narrative time, it has been argued, also engages the innovative experiments of modern literature, which also fractured and reordered linear time to convey its convoluted nature and experience at a viewer's will at subsequent multiple coordinates in space and time. This term from literary studies similarly provides the vocabulary for addressing the possibilities of an imaginative, subjective, simultaneous, multiple-tense photographic time. These ideas were further examined in the concluding section of Chapter 3. In an oral history for portrait of himself as a boy, Samuel B. Leighton recalls the Liberty Bell – an object that curiously finds no presence in the photograph. This case study implicitly argues that our subjective *narrative times* surpass the documentary power of photographs to testify to the appearance or presence of subjects. That is to say, our drive to temporalize such photographs often challenges our attachments to the images' veracity. The imposition of a subjective, creative, often non-chronological, narrative upon a photograph by a viewer also convolutes distinctions between *time* and *temporality*, implying that time's basis in human "reality" lies in its *embodiment* as the temporal. This phenomenon, Crary has argued, is a defining characteristic of visual habitude in an age of modernity.

Having examined the temporal complexities of photographic practices that enjoy a normative, fraction-of-a-second-exposure-time duration, Chapter 4 explores another means by which the medium expresses time's complex experience. Strip photographs – as well as others that are produced by a camera with a modified focal plane, shutter, and/or film advance system – defy the usual expectation that a photograph captures the appearance of a subject in a fixed, rectangular field of space at an atomized instant of time. They offer a perplexing collision of the durée and the millisecond, while representing racing subjects with a confounding contradiction of absolute realism and blatant abstraction. Such images suggest that the photograph's status as a truthful document is perhaps conditional. Chapter 4's analysis of the ubiquitous, yet understudied, racing photo finish – as well as its adjacent photographic practices such as panoramic, Cirkut, and peripheral/"rollout" photography, which also alter the recording field and exposure time – examines the medium's capacity to represent time and space as asynchronized, accumulated, distended, and/or compressed. In the photo finish, the time/space matrix is creatively reimagined and expressed.

Images such as the photo finish express an attempt to master time's relentless passage by capturing its constituent atomized bits with a sensitivity that exceeds unaided vision. As logical scientific armatures of the reason and positivist advance lauded by capitalist and industrial modernity, images such as the photo finish were hailed as human-made creations that possessed the means to bypass the human "failings" of subjectivity and restore a stabilizing force of objectivity and logic in an age in which rapid technological advance compressed, complicated, and reconfigured perceptions of space and time. However, their blatant distortions negate this sense of "mastery" and reaffirm the "other-ly" representations of time in photographs.

Chapter 5 introduced another unstudied photographic practice that allowed for the close examination of an added range of temporal expressions, NASA's whole-Earth digital composite images. Whole-Earth digital photographs feature a complex synthesis of atomized, differentially deferred times, extended duration, collapsed space, and simultaneity of vision. Time is distended, layered, mathematically abstracted, montaged together, compressed, and digitally repackaged to form a coherent, new, idealized, imaginary time.

Whole-Earth composite montages also appear frequently in our textbooks. However, they represent a digital tapestry of multiple, layered, simultaneous times, and reveal the Earth as it *never appeared.* These images are *real* without being *actual, symbols* rather than *evidences.* As such, they reveal the contradictions of photographic realism, and suggest the tandem nature of both temporal and veristic complexity. Whole-Earth digital composite photographs are temporal data visualizations that both enjoy the realist rhetoric of photography and the computational accuracy of resulting from its genesis as computer-generated data. Yet whole-Earth images offer an "unreal," an impossible and confounding collision of countless "atomized" views of time. These photographs also are products of an anxious culture coming to grips with a quickening pace of life enabled by the evolving, ever-*becoming* technology used to create and compose such photographic images.

Photography as Agent and Catalyst of Technological Modernity and Postmodernity

As the product, embodiment and faithful servant of a complex and variegated technological/industrial modernity and postmodernity, photography is equally variously temporally stratified. It provides a distinct means for expressing what it means to live in times of exhilarating industrial, scientific, and technological advances, and is a medium for productively channeling the resulting anxieties it stirs over the unfixed nature of temporality and reality. It is an ideal medium with which to express the temporal sensation of living in a relative world with no absolutes, operating without fixed frames of reference, where objects are dependent upon observing subjects to give them interpretive form and meaning.

Photography was born alongside inventions such as the passenger train and telegraph, which were widely credited with "annihilating time and space" by presenting to the general public the undeniable possibility of their instabilities. Within a decade of the announcement of the invention of photography in 1839, the passenger train (1830), computer (1833), and transatlantic telegraph (1844) were introduced, followed by the invention of the telephone (1876), automobile (1890s), cinema (1894), radio (1900–1910), airplane (1903), television (1939), internet (1969), the first popular personal computer (1976), and cell phones (1982).[2] This list is certainly an incomplete and ongoing one. This ongoing rush of technological advance accelerated the perceived pace of life dramatically, forever altering experiences and conceptions of space and time, absence and presence. As such, the medium inextricably has been bound to the malleable temporal experiences of industrial/technological modernity and postmodernity.

In particular, the innovations that emerged within a decade of photography's announcement introduced profound ruptures to sensory cognition. Technology, including photography, enabled what Oliver Wendell Holmes described as a revolutionary, perception-shaping "divorce of form and substance," as images functioned as substitutes for their referents. Circular grains (and, later, pixels) gave photographic subjects flattened external visual manifestations that functioned apart from their physical presences.[3] The visual translation of dots and dashes divorced language from its conventional form so that it could travel vast distances instantaneously and would later be reconstituted as language. Benjamin commented that photography has a peculiar capacity to reconstitute the presence of absent subjects, just as Holmes suggests telegraphy did for language. Benjamin elaborated that:

> Subjects of photography, seized by the camera, we are mortified – that is, objectified, thingified, imaged. The procedure itself caused the models to live, not out of the instant, but into it; during the long exposure they developed, as it were, into the image. The image already announces its absence.[4]

To Benjamin, the photograph had the peculiar capacity to create objects from subjects, and to transport them for re-viewing across infinite points in the time/space nexus. Being "thingified" or becoming "imaged" has ramifications upon the subject's presumed singular existence, as Benjamin suggests.

In addition to introducing a rupture or division between "form" and "substance," other commentators have noted that innovations of the 1830s created an "annihilation of space and time." On September 15, 1830, the day that the first passenger train line opened – with her aboard – actress Fanny Kemble marveled at the unprecedented speed with which the train covered the landscape. She commented that her mother was "frightened to death" to ride the train, considering it "a situation which appeared to her to threaten with instant annihilation herself and all her traveling companions."[5] In 1839, Ulysses S. Grant also recalled with amazement a trip he had made by rail that same year: "We traveled at least eighteen miles per hour when at full speed, and made the whole distance averaging as much as twelve miles an hour. This seemed like annihilating space."[6] Similar comments appeared in the *Baltimore Patriot* on the day after Samuel F. B. Morse's 1844 demonstration of the "Electro Magnetic Telegraph" at the United States Capitol. That article declared Morse's invention a "truly astonishing contrivance," and continued, "This is indeed the annihilation of space."[7] Ralph Waldo Emerson similarly wrote of the railroad's "annihilation" of distance, but worried that such a thing might bring the death of regional cultures.[8]

The phrase "annihilation of space and time" still is used today. Dorthe Gert Simonsen argues that:

> Phrases such as "conquering," "shrinking," and "annihilation" of space and time have been used in popular discourse to capture the experience of more than a century of accelerating technology. From this perspective "time-space compression," which

has been among the tropes most often repeated in the rhetoric of globalization and the Information Age, begins to sound like an echo of the former period.[9]

Since the announcement of its invention in 1839, photography has insistently struggled to control or reorganize time's passage, to halt or slow it for a moment, and to give form to time's destabilized conditions. Although many scholars such as Braun and Doane have characterized photography's intentions as a desire to rationalize or control time by giving it concrete form, the medium also celebrates time's subjective experience just as time itself was standardized and homogenized. Until the passenger train was invented, every town kept its own time, and accuracy and rigidity were not essential.[10] This ended with the introduction of standardized "railroad time," which was said to have "annihilated space and time," as Ierley argues:

> Before the era of the railroads, the "correct" time was almost anything approximating the hour of the day. In small towns and villages, and especially in the country, one's daily events were scheduled by the sun ... Most activities – working, farming, doing household chores – were not dependent on starting and stopping at a certain time. And an activity that was so dictated – say, going to church – could be time-coordinated for all involved by the ringing of a bell.[11]

This reactive urge to quantify, manage, and measure time after its standardization quickly found its way into modern industry. In 1844, Charles Babbage (the "father of computing" and a key member of the circle of the British intelligentsia that included pioneer of early photography, Talbot) invented a mechanical clock that was used in workplaces to measure and monitor the length of employees' work shifts.[12] The very consciousness of time's value is itself a product of an age of modernity and industrialization. Managing "lost" time was an especially pressing concern in the twentieth century, after Frank and Lillian Gilbreth and Frederick Winslow Taylor demonstrated (with the visual aid of extended-exposure photographs of workers wearing light probes on their hands) that workplaces could increase productivity by eliminating excess human motion in repetitive tasks. As a result, the early to late twentieth century

> was a time when people found themselves pressured by the clock. Experts in "scientific management" segmented and streamlined both factory and office work to increase productivity and advocated time-saving efficiencies at home. Even periods of "time off" became defined by the clock. Time was divided up and measured out, not to be wasted ... [T]ime could be turned into money: wages for workers and into products and profits for the employer. Managers began saving time by measuring and controlling the amount of time a worker spent on any task. Efficiency became a manager's expectation.[13]

"Scientific management" was immediately popular because it capitalized on the new twentieth-century obsession with speed, setting records, and trying to "beat" time. These endeavors were widely celebrated in the mass media. For instance, before it tragically collided with an iceberg and sank, the HMS *Titanic* sought to set a speed record and save an unprecedented amount of travel time.[14]

In his study of everyday life in urban centers, Georg Simmel argues that simultaneous, fast-paced, competing visual stimuli in an age of modernity prompt an "intensification of nervous stimulation" which traumatizes, reorders, and makes observers' perceptions of time subjective.[15] He writes that the metropolis specifically creates a psychological climate in which fleeting momentary relationships reconfigure vision, and strain the nervous system:

> Impressions which take a regular and habitual course and show regular and habitual contrasts – all these use up, so to speak, less consciousness than does the rapid crowding of changing images, the sharp discontinuity in the grasp of a single glance and the unexpectedness of onrushing impressions.[16]

According to Simmel, the human capacity to accommodate such a "rapid crowding" of stimuli was tested in an age of modernity. Likewise, writing in 1901, historian John Girdner also identified what he saw as a new social disease afflicting all urban-dwellers, "Newyorkitis." This uniquely modern psychological condition caused a "rapidity and nervousness and a lack of deliberation in all movements," and was caused by modern stimuli's high-speed, multi-layered assault on a previously linear, slower-paced perception of the pre-modern age.[17] In a medical study, Willy Hellpach specifically tied this new state of "nervousness" to the increasing rapidity of communication and transportation in the nineteenth century, which caused an "overwhelming increase of normal mental processes."[18]

The embodiment and interpretation of time/space perceptions by the observer, art historian Jonathan Crary argues, renders temporality unfixed, subjective, and contingent in an age of industrial modernity – and particularly in the mid- to late nineteenth century, as photography popularity became increasingly widespread.[19] Around the fin de siècle, Eastman Kodak's introduction of a relatively affordable, amateur-friendly box camera with an exposure duration of 1/20th of a second to the mass general public made the medium's presence yet more visible, and yet more engaged in the reconfiguration of human temporal perception. In particular, it fully brought photography into the fold of mass visual culture.

Public pushback did occur. In his accounts on the growth of the Trans-Siberian railroad line, Octave Uzanne writes that, although the public expressed a nostalgia for the rhythmic clip-clop of horses' hooves along the road, the railroad also brought a "stupefying transfiguration," in which "rapid movement which sweeps us up in space and piles up a variety of impressions and images in a short time gives life a plentitude and a unique intensity."[20]

Indeed, the new brisk speed of travel redefined space and time, and the body's experience while moving at a new and foreign speed through it. As one passenger commented in South Carolina's *Charleston Courier* in 1830: "We flew on the wings of the wind at the varied speed of fifteen to twenty-five miles an hour, annihilating time and space."[21] In the 1830s, a walking pace of two to three miles an hour was as fast as most people ever passed through the landscape, and speeds of about 55 miles per hour induced nausea.[22] In the 1830s, passengers in Philadelphia could reach Lancaster in two hours, Pittsburgh in ten hours, and New Orleans in fewer than two days, and could return to Philadelphia within a week.[23] As a result, the modern age saw a redefinition of time and space as unstable entities, as Dorthe Gert Simonsen argues, but this condition merely accelerates in the postmodern age:

> If the time needed to traverse space units can be endlessly reduced through the use of mechanical means of transport, then not only does the relation between time and space become fluid but also the categories themselves disintegrate. In other words, time and space were torn out of their Newtonian stability during the modern period.[24]

By destabilizing conventional notions of time and space, the passenger train's invention ushered in an age of time-consciousness and time anxiety that persisted into the twentieth century, and up to today. Inventions such as the passenger train had a profound effect upon an embodied sense of modern temporality. Filippo Tommaso Marinetti, spokesman of Italian Futurism, asserted:

> Those people who today make use of the telegraph, the telephone, the phonograph, the train, the bicycle, the motor cycle, the automobile, the ocean liner, the dirigible, the aeroplane [sic], the cinema, the great newspaper ... do not realize that these various means of communication, transportation and information have a decisive influence on their psyches.[25]

This fixation on speed, "efficiency," and ever-more-mindful time management brought dangerous consequences. "Scientific management" enhanced power inequities between workers and managers, while giving modern industry the illusion of having "conquered" time by showing them a way to take advantage of every split second of productivity. In addition, efficiency solutions often forced workers to perform the job so quickly that they put their lives in danger. For example, after Taylor implemented changes to improve workplace efficiency at Bethlehem Steel, his standards "made the work rates so difficult that only one worker out of seven could perform them."[26] As the workday moved at a warp-speed pace, the temporal experience of a day at the workplace was intense and hyper-stimulating. Employees felt fatigued, thanks to this new science and technology and to the new "valuing" of minute increments of time in an age of industrial modernity.

Although managers and business owners sang the praises of "scientific management," the working class soon felt its enthusiasm for this new, positivist, time-saving "science" turn to trepidation and anxiety.[27] Magazines lampooned the idea of efficiency and grace in everyday movements. But the movie *Modern Times* (1936), for instance, reflects a darker view of technology and time, brought on by "scientific management."[28] As actor Charlie Chaplin worked to tighten bolts with a single, machine-like motion on an assembly line, his boss kept speeding up the line. Chaplin finally found solace in an asylum, where he was free at last from pressing and unrealistic productivity quotas.[29] In an episode of *I Love Lucy*, Lucille Ball lost her job in a candy factory because she could not keep pace with an assembly line – even after eating several of the candies in a vain effort to lighten her workload.[30] The popular song "Racing with the Clock," for example, also complained that workers were competing against the clocks, which would always have the winning edge.[31]

As the twentieth century gave eventual rise to the personal computer, cell phones, and the internet, the gap between the bodily rhythms of time and the technology-encouraged one declined precipitously. Time has become an even more pressing issue in technological postmodernity. Doane notes: "The lure of the internet is the lure of connectivity, of being in touch, of synchronicity, and of availability – 24/7/365: 24

hours a day, 7 days a week, 365 days a year."[32] The internet's temporal cadence never stops. It is relentless, as our hand-held and laptop devices challenge us to keep up. As a whole, Timothy Druckrey suggests:

> It is no exaggeration to suggest that the cumulative effects of science and technology on twentieth century culture have challenged, reoriented, triumphed over, exploded, and infiltrated into notions of the self, the mechanisms of the biosphere, and the analysis of matter. The entire concept of communication has likewise been extended into previously unimagined realms. And, as the [twentieth] century closes, the daunting significance of computation is extending the material sciences into speculative, genetic, artificial, and rendered territories, inconceivable just a few decades ago.[33]

Time anxiety found expression in not only these movies, but in other areas of mass culture – including photography. The assimilation of technology's new temporal processes cannot be fully accommodated immediately. As early as the 1920s, technology historian William F. Ogburn explained that "culture takes time to catch up with technological processes."[34] Specifically, in the ages of technological modernity and postmodernity, viewers needed time to absorb the new, increasingly structured, regulated, and rapid, pace of time. But there was *no time* to reflect on such changes. Scholar Charlie Gere explains the anxieties inducing this "cultural lag" as prompted by "confronting limits beyond which the structures through which we have traditionally understood the world no longer operate."[35] Worry over "losing" valuable, previously unnoticeable, time in the workplace (combined with the loss of the familiar past to modernity's rush of newness) created a positivist urge to harness science and technology to capture every moment of modern and postmodern life, even the infinitesimal ones.

One reaction to the anxiety over the increasing speed of stimuli was to redefine time, in order to exert some degree of control over it. As a result, "[t]ime has been redefined as an ecological niche to be filled down to the microsecond, nanosecond, and picoseconds," James Carey notes. Yet, by creating ever smaller units of labeled (and therefore, *defined*) time, its units are reduced to intangible abstractions.[36] Order dissolves into chaos. Photography has been one means by which humankind has grappled with time's strangeness. The medium's unique, innately estranged views of time can express the unfixed experience and sensations of time's passage in fluid visual form.

Time Embodied as Temporality, Subjectivity

An embrace of temporality's contingent, subjectively perceived, and unstructured nature provided an alternative to the regimented nature of standardized time. Many various philosophers and scientists – including Umberto Eco, Martin Heidegger, Henri Bergson, and Charles Sanders Peirce – theorized the experience of temporal subjectivity. As Jan Schall notes, the mass public has absorbed and internalized these advances:

> We now live in a simultaneous world, experiencing distant events through real-time transmissions in the privacy of our homes and offices. We have absorbed

> the sense, if not the particulars, of Albert Einstein's special and general theories of relativity, which established space-time as a unified continuum. Henri Bergson's philosophical treatises, which defined time not as a quantifiable entity but as a subjective experience of duration, have likewise been assimilated. Sigmund Freud's groundbreaking research into the significance of the fluid, nonsequential time we experience in unconscious and dream states has also been internalized.[37]

The terms typically used to describe and spatialize one's position in linear time – *past, present,* and *future* – are themselves unfixed and often inadequate. For instance, Umberto Eco argues that time only exists as *past* tense. He suggests that our awareness of *the present* is actually of events that have *passed* (and therefore already are in *the past*).[38] Eco further argues that we have no experience with which we can attest to the existence of *the future*.[39] And in the 1927 book *Being and Time*, the German philosopher Martin Heidegger suggested an alternative to the conception of time as existing as successive phases of the past, present, and future. He argued that the past impresses itself upon the present, which is shaped by expectations of future events. Thus, interpenetrating tenses of past, present, and future inform each other and create the temporal dimension of human existence.[40] This convoluted experience of time, Heidegger argues, comprises our very state of *being*.

Henri Bergson suggested in his essay *Time and Free Will: An Essay on the Immediate Data of Consciousness* that time's nature is rooted in its innate subjectivity, and it lies beyond the capacity of understanding offered by systemized, mathematical management. He postulated that "lived time" is experienced as *la durée* (duration), and it "is psychical in its nature and psychological in its order . . . [and] is non-quantitative, and therefore non-measurable and indivisible."[41] Bergson's definition of "extensity" also posed the possibility that our sensory experiences necessarily embody intermingling spaces and times.[42] As such, our realities are also subject to change, Bergson suggests, and also are subject to individual interpretation and intuition.[43] Bergson also envisioned time as an agent for enabling the process of change, or "becoming." The photograph's ability to still the durational "*-ing*," or "becoming," of a subject – as illustrated in the extended exposures of Daguerre's *Boulevard du Temple* (Fig. 2.4) – exemplifies Bergson's concept of duration, while also illustrating the literal act of a subject's "becoming" an image on the photographic plate. As Charles Sanders Peirce suggests, every photograph embodies duration: "[e]ven what is called an 'instantaneous photograph,' taken with a camera, is a composite of the effects of intervals of exposure more numerous by far than the sands of the sea."[44] Bergson's theory of temporal contingency gained scientific affirmation in the theories of relativity offered by Albert Einstein, which proved that the hands of clocks aboard trains moved faster than clocks that were kept on still ground. Einstein's theories posed the possibility that the fourth dimension's spatial and temporal coordinates are fluid, and time's passage therefore is unfixed, mutable, and subjective.[45] Relative-time theories by physicists such as Einstein worked to steadily dismantle the Newtonian conceptions of an absolute, fixed time.[46] As the authority of a standardized time was imposed worldwide in an age of modernity, individuals – particularly inspired by these highly publicized advances in physics – embraced its personal, subjective experience while also noting the inadequacy of various equally abstract ordering systems to control or manage the intangible entity of time.[47]

Albert Einstein's groundbreaking 1905 "Special Theory of Relativity" gave subjective perceptions of temporality scientific credibility.[48] By comparing the same duration of time quantified by a still clock to that measured by a clock traveling inside of a train, Einstein demonstrated that time's passage – or at least, its quantification using mechanical devices – was subjective. A moving clock measured the same duration of time more slowly than a stationary one. Einstein's "Theory of Relativity" of 1916 expanded this earlier theory and concluded with the statement that "every reference body has its own particular time."[49] In effect, Einstein disproved the existence of an absolute time, despite the rigid imposition of standardized measurement systems and the widespread implementation of "unbiased" mechanical devices.

Philosopher Martin Heidegger also offered an alternative to the conception of time as a linear order of the three distinct tenses of past, present, and future. He argued that these tenses are "merely arbitrary linguistic notations for qualitatively similar segments of a continuous series of measurable bits."[50] Heidegger argued that the past constantly shapes and infringes upon the present, which also is shaped by the future. Heidegger – as well as phenomenology's founding father, Edmund Husserl – theorizes a time in which past, present, and future interpenetrate to form the core of human existence, temporality.[51]

Theories about the increasingly subjective nature of temporality by commentators such as Benjamin, Bergson, Einstein, and Heidegger led to an anxiety that too much temporal subjectivity had led to the unfixing of the notion of absolute knowledge. Specifically, it was feared that "the notion of knowledge as accurate representation … and intelligible through a general theory of representation, needs to be abandoned," as Richard Rorty argues.[52] Yet as theories of temporal subjectivity enjoyed traction following World War II, German sociologist Arnold Gehlen attributed the unfixing of absolute knowledge to technological progress – which he argued outpaced human evolution, and led modernized societies into a state of "post-history."[53] A state of "post-history," as Charlie Gere explains, is an age characterized by "a sense of confronting limits beyond which the structures through which we have traditionally understood the world no longer operate."[54] It therefore is an age marked by anxiety. Specifically, Gere argues that these fears have manifested themselves in apocalyptic predictions of nuclear disaster, prophesies about ecological catastrophe and overpopulation, forecasts of the end of humanity, and the end of history, grand narratives, philosophy, religion, and scientific advance.[55]

These anxieties also permeate an art-historical discourse. Painting, photography, and art in general have all been declared "dead" – or radically, essentially redefined as media – by various commentators from the time of photography's invention to today.[56] Photography's peculiar capacity to provide a materialized, reconstituted, two-dimensional presence of a three-dimensional object as it appeared in past time makes the medium an ideal site for exploring contingency anxiety, and temporal convolution.

These commentators' efforts to give scientific and philosophical grounding to subjectivity – in order to systemize and find uniquely "modern" grounding for a fear of ephemerality – reflect broader cultural anxieties toward technology's distortion of temporal perceptions. Crary has argued that modernity has been marked by the observer's embodiment of subjectivity.[57] The general public – the audience of the photographic reproductions about which Benjamin theorizes – expressed deep ambivalence about technological advance in an age of modernity. While some people "looked forward to happier utopias with less drudgery, cheaper goods, and clean, safe cities, others saw mixtures of progress and degeneration, islands of care-free pleasure and oppressive technologies," according to Stephen Kern.[58]

As a consequence, these various commentators noted that time itself has been the subject of insistent theorization, speculation, and anxiety in an age of modernity. Their writings locate modernity's temporality in the birth of individual subjectivity. Crary grants importance to the role of the viewer in interpreting and determining reality – which Crary argues is shaped by an innate, embodied human subjectivity unique to the age of modernity.[59] Although Baudelaire suggested that an image capturing a fleeting moment of time encapsulates a decisively "modern," fragmented experience of temporality, Crary complicates this view by arguing that a viewer's subjective, changeable *temporal processing* of such pictures is the essence of their "modernity." These commentators suggest that a "modern time" is redefined as an embodied, individualized "temporality." But I also would add that "modernity" – a phenomenon not synonymous with "Modernism," although many Modern artists deal with it as a subject – is an age of simultaneous, competing, compressed, and increasingly subjective, times and spaces, and in an age in which time's standardization has conflicted with its biological, personalized, embodied experience.

The invention of photography, perhaps not entirely without coincidence, emerged as Newtonian postulates of concrete, steady temporal and spatial coordinates were dissolved by the new scientific theories of quantum mechanics of many key figures, including Einstein, Boris Podolsky, and Nathan Rosen. The unfixing of the matrices of the known world likewise relegated space, time, and matter to provisional status. Theories of temporal subjectivity by Bergson and Heidegger (and others) further worked to encourage this view of a relative world with no absolutes, operating without fixed frames of reference, where objects are dependent upon observing subjects to give them interpretive form and meaning.

Photography's expressions of temporality are many, are complex and layered. Time in a photograph is not *embalmed*, as Bazín suggests, but it is sculptural, multiplicitous, dynamic, and *enlivened* by the viewer. Moreover, photography becomes a unique medium for expressing the sensation of living with time's unfixity and relentless, relentless passage. It fittingly expresses the juxtaposition of our humble synthetic systems and mechanisms against the task of understanding larger forces and flows, such as time.

In closing, it is helpful to return for a moment to the words of St. Augustine: "Who shall hold [time], and fix it, that it be settled awhile, and awhile catch the glory of that everfixed Eternity, and compare it with the times which are never fixed, and see that it cannot be compared."[60]

It is worth noting the similarity between St. Augustine's words and those of Talbot, one of photography's founders, as he describes the unique capacity of his early photographic experiments to access eternity by "fixing" a moment:

> The most transitory of things, a shadow, the proverbial emblem of all that is fleeting and momentary, may be fettered by the spells of our "natural magic," and may be fixed forever in the position which it seemed only destined for a single instant to occupy.[61]

St. Augustine above humbly argues that only God is beyond time and is capable of understanding its complex operations. Talbot, whose new innovation compellingly can coax the Godly force of time to be persuaded by humankind's "natural magic" and appear as "the most transitory of things, a shadow" That is to say, photography, Talbot implicitly argues, rivals God's access to time.

Such a vital innovation held promise for expressing of what it means to live in a time of modernity – with all of its exhilaration, scientific, and technological advance, and the resulting anxieties it stirs over the unfixed nature of temporality and reality. It is an ideal medium with which to express the temporal sensation of living in a relative world with no absolutes, operating without fixed frames of reference, in which objects are dependent upon observing subjects to give them interpretive form and meaning.

Notes

1 Olin, "Touching Photographs : Roland Barthes's 'Mistaken' Identification." *Representations*, Vol. 80 (Fall 2002): 105, doi: 10.1525/rep.2002.80.1.99; Roland Barthes, *Camera Lucida: Reflections on Photography* (New York: Hill and Wang, 1981): 87–88.

2 Merritt Ierley, *Wondrous Contrivances: Technology at the Threshold* (New York: Clarkson Potter Publishers, 2002).

3 Oliver Wendell Holmes is quoted in Beaumont Newhall, ed., *Photography: Essays and Images* (London: Secker and Warburg, 1980), 59–60. Geoffrey Batchen notes the similarities between the Holmes's observation of "the divorce of form and substance" in the nineteenth century and today's cyberculture in this essay: Geoffrey Batchen, "Electricity Made Visible," in *New Media, Old Media: A History and Theory Reader*, Wendy Hui Kyong Chun and Thomas W. Keenan, eds. (New York: Routledge, 2006), 33.

4 Benjamin, "Little History of Photography," 245.

5 Rebecca Solnit, *River of Shadows: Eadweard Muybridge and the Technological Wild West* (Middlesex, England: Viking Penguin, 2003), 9.

6 Solnit, *River of Shadows*, 9.

7 Ierley, *Wondrous Contrivances*, 5.

8 Solnit, *River of Shadows*, 10–11.

9 Dorthe Gert Simonsen, "Accelerating Modernity: Time-Space Compression in the Wake of the Aeroplane," *Journal of Transport History* Vol. 26, No. 2 (2005): 102.

10 Doane, *The Emergence of Cinematic Time: Modernity, Contingency, the Archive*. Cambridge, MA, and London: Harvard University Press, 2002, 11. Starting with the standardization of railroad time, "Clocks began dominating every arena of daily life, from the fields to factories, from school and work to home"; National Museum of American History – The Smithsonian Institution, "'On Time' Opens at Smithsonian's National Museum of American History," available at Smithsonian http://americanhistory.si.edu/news/pressrelease.cfm?key=29&newskey=97 (accessed April 22, 2007). No longer did a village bell ring to bring churchgoers to worship services, or to signal the start of the work day. As a result, people "either fell easily into the pattern and conveniences of standard time or resented its rigid order. This was a time of waking to the jangle of an alarm clock on the bedside rather than to the sun. The clock determined the patterns of home and work rather than nature's intrinsic rhythms"; National Museum of American History – The Smithsonian Institution, "'On Time' Opens at Smithsonian's National Museum of American History," n.p.

11 Ierley, *Wondrous Contrivances*, 29.

12 Jo Ellen Barnett, *Time's Pendulum: From Sundials to Atomic Clocks, the Fascinating History of Timekeeping and How Our Discoveries Changed the World* (New York and London: Harcourt Brace and Company 1998), 141.

13 National Museum of American History – The Smithsonian Institution, "'On Time' Opens at Smithsonian's National Museum of American History," n.p. "Scientific management"

was well-intended: "The Chicago World's Fair in 1933 unveiled the Gilbreth Management Desk, which was promoted as the 'Business Headquarters of the Household Manager.' The desk, which was intended for the kitchen, has a clock within easy reach, a radio, a telephone, and adding machine, household files, reference books, schedules, and a series of pull-out charts with tips on organizing and planning household tasks. Dr. Lillian Gilbreth, an industrial psychologist and a pioneer in factory efficiency studies and the mother of twelve children, saw the home as a workplace and the homemaker as both worker and manager. Her goal was to increase productivity, reduce drudgery, and accumulate the time saved for leisure or creative pursuits"; National Museum of American History – The Smithsonian Institution, "'On Time' Opens at Smithsonian's National Museum of American History," n.p.

14 Mike Mandel, *Making Good Time: Scientific Management –The Gilbreths, Photography and Motion, Futurism* (Riverside, CA: Museum of Photography, University of California, 1989), 10.
15 Georg Simmel, "The Metropolis and Mental Life," in *The Sociology of Georg Simmel*, Kurt H. Wolff, ed. (New York: Free Press, 1950), 410.
16 Simmel, "The Metropolis and Mental Life," 410.
17 John H. Girdner, *Newyorkitis* (New York: 1901), 119.
18 Willy Hellpach, *Nervousness and Culture* (Berlin: 1902), 12.
19 Crary, *Techniques of the Observer: On Vision and Modernity in the Nineteenth Century.* (Cambridge, MA: MIT Press, 1990).
20 Octave Uzanne, *La Locomotion à Travers le Temps, les Moeurs et l'Espace* (Paris: 1912), vi–vii, 244–247, 304. Uzanne is quoted in Stephen Kern, *The Culture of Time and Space, 1880–1918* (Cambridge: Harvard University Press, 2003), 128.
21 Ierley, *Wondrous Contrivances*, 7–8.
22 Ierley, *Wondrous Contrivances*, 18. Before train travel was invented, some boats were capable of transporting people at an average of 1.5 to 5 miles per hour. Stagecoaches were the fastest land-based travel before the railroads emerged. The average top speed for a stagecoach was 6 to 8 miles per hour. One route between Philadelphia and New York City reached speeds up to 11 miles per hour, as the result of competition between coach lines. This was exceptionally fast travel for the time; Ierley, *Wondrous Contrivances*, 18.
23 Ierley, *Wondrous Contrivances*, 19.
24 Simonsen, "Accelerating Modernity," 101.
25 Simonsen, "Accelerating Modernity," 100–101. This quote appears in: Filippo Tommaso Marinetti, "Destruction of Syntax – Imagination Without Strings – Words in Freedom (1913)," in *Futurist Manifestoes*, Umbro Apollonio, ed. (Boston: Viking Adult, 2001), 96.
26 Mandel, *Making Good Time*, 9.
27 Positivism and photography are often seen as having common roots: "[T]he camera was invented in 1839. Auguste Comte was just finishing his Course de Philosophie Positive. Positivism and the camera … grew up together"; John Berger, *Another Way of Telling* (London and New York: Vintage International reprint/Penguin-Random House, 1995), 99.
28 Charlie Chaplin, *Modern Times* (Charlie Chaplin Productions, 1936). In 1865, the British writer and mathematician Lewis Carroll (Charles Lutwidge Dodgson) published the book *Alice's Adventures in Wonderland*, which featured these famous lines that indicate time awareness: "I'm late / I'm late / for a very important date!" Lewis Carroll, *Alice's Adventures in Wonderland* (London: Macmillan and Co., 1865).
29 Chaplin, *Modern Times.*
30 Desi Arnaz, "Job Switching/Candy Factory," in *I Love Lucy*, Season 2, Episode 39 (1952).
31 Jerry Ross and Richard Adler, "Racing With the Clock," in *The Pajama Game* (1954).
32 Mary Ann Doane, "Real Time: Instantaneity and the Photographic Imaginary," in *Stillness and Time: Photography and the Moving Image*, David Green and Joanna Lowry, eds. (Brighton: Photoworks/Photoforum, 2006), 24.
33 Timothy Druckrey, *Electronic Culture: Technology and Visual Representation* (New York: Aperture Foundation, 1996), 13.
34 Langdon Winner, *Autonomous Technology: Technics-out-of-Control as a Theme in Political Thought* (Cambridge, MA: MIT Press, 1977), 77. Although it was generally

acknowledged that such a lag existed, Charlie Gere suggests that this phenomenon became most obvious during World War II, due in part to the appearance of computing: "The development during and after the War of radar, jet propulsion, digital computing and atomic weaponry presented vivid proof of the speed and reach of technological evolution"; Charlie Gere, *Art, Time and Technology*, Culture Machine Series (Oxford and New York: Berg, 2006), 32.

35 Gere, *Art, Time and Technology*, 33.

36 James W. Carey, *Communication as Culture: Essays on Media and Society* (Boston: Unwin Hyman, 1989), 228–229. Carey also argues that the telegraph revolutionized commerce, especially in America, because it regularized time: "the telegraph, via the grid of time, coordinated the industrial nation"; Carey, *Communication as Culture*, 229. The telegraph, along with the invention of rail travel, allowed people to do business with people they did not know personally. This was a real boost to American commerce in the nineteenth Century. Carey, *Communication as Culture*, 205. Telegraph inventor Samuel F. B. Morse foresaw the impact of the telegraph. He anticipated that it would have the power to create a "global village," able to share thoughts with each other instantaneously. It would not be long, Morse wrote, "ere the whole surface of this country would be channeled for those nerves which are to diffuse with the speed of thought, a knowledge of all that is occurring throughout the land; making in fact one neighborhood of the whole country"; Carey, *Communication as Culture*, 207. The rhetoric surrounding the invention of the telegraph hailed this new invention as the source of enlightenment and progress, suggesting that "isolation and disconnection are evidence of barbarism and merely obstacles to overcome . . .Each improvement in communication, by ending isolation, by linking people everywhere, was heralded as realizing the Universal Brotherhood of Universal Man"; Carey, *Communication as Culture*, 208.

37 Jan Schall, *Tempus Fugit: Time Flies* (Kansas City, MO: Nelson-Atkins Museum of Art, 2000), 14. John House further suggests that, "By the mid-nineteenth century, time had become a central issue in wholly new ways in the lives of populations across the Western world. In everyday life, the speed of railway transport transformed the experience of the immediate environment. Instead of the gradually shifting views seen from horse-drawn transport, the railway passenger was offered a kaleidoscopic sequence of new and unexpected perspectives; and the dramatically shortened travel times seemed to make the world a far smaller and more accessible place. In broader terms, too, the increasing rapidity of change – technological, industrial, social, economic, political – meant that for the first time radical alterations in world view were to be anticipated within a single lifetime. The inhabitants of the modern city could no longer expect to pass on to their children what they themselves had inherited from their parents. And, in cosmic terms, the discovery of 'deep time' revolutionized the understanding of man's place in the universe, at the same time as challenging any literal interpretation of the Judeo-Christian narrative of creation"; John House, "Seasons and Moments: Time and Nineteenth-Century Art," in *The Story of Time*, Kristen Lippincott, ed. (London: Merrell Holberton, 1999), 194.

38 Umberto Eco, "Times," in *The Story of Time* (London: Merrell Holberton, 1999), 10–15.

39 Eco, "Times," 10–15.

40 Martin Heidegger, *Being and Time* (1927), Joan Stambaugh, trans. (Albany: State University of New York Press, 1996).

41 Henri Bergson, *Time and Free Will: An Essay on the Immediate Data of Consciousness*, F. L. Pogson, trans. (London: George Allen and Unwin, 1910), 105–106, 115–116, 120–121.

42 Henri Bergson, *Matter and Memory (1896)*, Nancy Margaret Paul and W. Scott Palmer, trans. (London: George Allen and Unwin, 1911), 92–93, 326. "The extensive," according to Bergson, lies between "divided extension," or abstract Euclidean space, and "pure inextension," or duration. Bergson, *Matter and Memory*, 92–93. The Futurists differ from Bergson on the possibility and potential of simultaneity. The Futurists (including the Bragaglia brothers) use multiple simultaneous visual stimuli to reflect their affinity for dynamic, relentless energy. In Anton Bragaglia's photographs, linear time is not discernible, and body parts are caught in a blurry trace of previous motion. Motion is indicated throughout the photographs, but without clear evidence of the

sequence. Without apparent linear "order" for defining the priority sequence of the physical events represented, these blurs and repeated forms are interpreted as simultaneous. Bergson, on the other hand, did not believe that simultaneity was possible, Nancy Ritcher argues; Nancy Ritcher, "Understanding Simultaneity in Italian Futurist Art," *Contrapposto* [Online journal, California State University at Chicago], available at www.csuchico.edu/art/contrapposto/contrapposto00/pages/themecontent/ritcher197.html (Accessed September 29, 2007).

43 Bergson, *Matter and Memory*, 92–93.

44 Charles Sanders Peirce, *Collected Papers of Charles Sanders Peirce*, Charles Hartstone and Paul Weiss, eds. (Cambridge: Harvard University Press, 1932), 267. Peirce's semiotic classification of the photograph as both an "icon" and an "index" allowed the image room to flip freely in and out of semiosis with its referent: Peirce, "Logic as Semiotic: The Theory of Signs." In *Semiotics: An Introductory Anthology*, Robert E. Innis, ed. (Bloomington: Indiana University Press, 1985), 1–23.

45 Albert Einstein, *Relativity: The Special and General Theory*, Robert Lawson, trans. (New York: Henry Holt, 1920). To read more about the influence of relativity upon art in the early twentieth century, see Thomas Mook and Delo E. Vargish, *Inside Modernism: Relativity Theory, Cubism, Narrative* (New Haven and London: Yale University Press, 1999).

46 Historian of science Jimena Canales writes that the 1921 volume of reports by France's Académie des Sciences was "filled with new essays about the redefinition of time proposed by Einstein." Specifically, Charles Richet examined the work of Bergson and Einstein in terms of developing his own theories about psychological dimensions of time: Jimena Canales, *A Tenth of a Second: A History* (Chicago: University of Chicago Press, 2009), 179; various authors, *Comptes rendus ses séances de l'Académie des sciences*, Vol. 173, No. 25 (1921).

47 Kern, *The Culture of Time and Space*, xii.

48 Albert Einstein, *Relativity: The Special and General Theory* (1916), Robert Lawson, trans. (New York: Henry Holt, 1952).

49 Einstein, *Relativity: The Special and General Theory*, 26. Interestingly, Stephen Kern notes: "Einstein, who could not afford to have a clock on the wall of his room when he was working in the patent office in Berne, had filled the universe with clocks each telling a different correct time"; Kern, *The Culture of Time and Space*, 19.

50 Stephen Kern, "Time and Art in Twentieth-Century Culture," in *Tempus Fugit: Time Flies*, Schall, ed., 25. Heidegger's theories of temporality can be found in Martin Heidegger, *Being and Time*.

51 Edmund Husserl, *The Phenomenology of Internal Time-Consciousness*, Martin Heidegger, ed., second edition (Bloomington, IN, and London: Indiana University Press, 1966).

52 Richard Rorty, *Philosophy and the Mirror of Nature* (Princeton: Princeton University Press, 1979), 6.

53 Arnold Gehlen, *Man in the Age of Technology* (New York: Columbia University Press, 1980).

54 Gere, *Art, Time and Technology*, 33.

55 Gere, *Art, Time and Technology*, 33.

56 French Academic painter Paul Delaroche famously (and debatably) declared upon viewing a daguerreotype that "from today, painting is dead!" Although this statement is often quoted and attributed to him, Delaroche – who was an *advocate* of photography – may not have said it. Nonetheless, this quote is often summoned to express the anxieties that photography's invention prompted. It was feared that photography might replace painting as a reigning form of representation. Photography's "death" (an event which has been given various names, including "post-photography," and the "post-medium condition") was declared when the invention of digital manipulation software undermined what some theorists considered a defining characteristic of the medium – a given and absolute degree of indexicality. Art has been declared "dead" by commentators such as Donald Kuspit, who has argued that contemporary art has lost its sacredness, its soul, and its creativity in favor of an anti-aesthetic that accentuates the banal and the clever: Donald Kuspit, *The End of Art* (Cambridge: Cambridge University Press, 2004). Arthur C. Danto also

suggests that art died in 1964 with the creation of Andy Warhol's Brillo box installations – which he argues challenged the definitions of art and non-art. Danto sees today's art as deviating from the art-historical narrative – and as such, it no longer is "art" as we know it: Arthur C. Danto, *After the End of Art* (Princeton: Princeton University Press, 1998). In this paper, I consciously use the term "modernity" rather than speak in the divisive and hotly debated art-historical terms of "Modernism" and "postmodernism." The nebulous term "Modernity," for the purpose of this project, is defined in Husserl's terms, as a protest against the notions that time and space are objective truths of the natural world: Edmund Husserl, *The Crisis of the European Sciences and Transcendental Phenomenology: An Introduction to Phenomenological Philosophy*, David Carr, trans. (Evanston: Northwestern University Press, 1970), 316. In his study of Husserl's theory of modernity, Anson Rabinbach places this observed time-and-space distortion in its cultural context: "Cultural modernity has generally been understood in terms of the intimate connection between the rapid industrialization of the second half of the nineteenth century and the corresponding disruption of traditional space-time perceptions in science, art, literature, and philosophy. The visual and narrative field was dissolved by avant-garde writers and artists from Proust to Picasso while the natural sciences and mathematics were being transformed by thermodynamics and the emergence of non-Euclidean geometries. This widely shared sense of a crisis of space and time found cogent philosophical expression in Henri Bergson's assault on the 'spatialized' time of 'timekeeping' and measurement;" Rabinbach, *The Human Motor*, 84. *Photography, Temporality, and Modernity: Time Warped* finds its parameters for exploring "modernity" in Rabinbach's and Husserl's above terms, and does not actively engage in debates about the territories and defining characteristics of the "Modern" and "postmodern" (which are a central project in the aforementioned books by Kuspit and Danto). In its efforts to articulate continuities between the "Modern" and "postmodern," the book in effect takes the position that the "postmodern" does not mark a radical break with the "Modern."

57 Crary, *Techniques of the Observer*.

58 Kern, *The Culture of Time and Space*, 98.

59 Crary, *Techniques of the Observer*.

60 St. Augustine of Hippo, *Confessions* (CE 397–398), in *Augustine: Confessions*, Albert C. Outler, ed. and trans. (Dallas: Southern Methodist University Press, 1955), Book 11, Chapter 11, Section 13.

61 William Henry Fox Talbot, "Some Account of the Art of Photographic Drawing, Or, the Process by Which Natural Objects May Be Made to Delineate Themselves Without the Aid of the Artist's Pencil" (1839), in *Photography: Essays and Images*, Beaumont Newhall, ed. (New York: The Museum of Modern Art, 1980), 25.

Bibliography

Barthes, Roland. *Camera Lucida: Reflections on Photography*. New York: Hill and Wang, 1981.

Batchen, Geoffrey. *Burning With Desire: The Conception of Photography*. Cambridge, MA: MIT Press, 1997.

Canales, Jimena. *A Tenth of a Second: A History*. Chicago, IL: University of Chicago Press, 2009.

Carey, James W. *Communication as Culture: Essays on Media and Society*. Boston: Unwin Hyman, 1989. 228–229.

Doane, Mary Ann. "Real Time: Instantaneity and the Photographic Imaginary." In *Stillness and Time: Photography and the Moving Image*. David Green and Joanna Lowry, eds. Brighton: Photoworks/Photoforum, 2006. 22–38.

Druckrey, Timothy. *Electronic Culture: Technology and Visual Representation*. New York: Aperture Foundation, 1996.

House, John. "Seasons and Moments: Time and Nineteenth-Century Art." In *The Story of Time*, Kristen Lippincott, ed. London: Merrell Holberton, 1999. 194–197.

Ierley, Merritt. *Wondrous Contrivances: Technology at the Threshold*. New York: Clarkson Potter Publishers, 2002.
Kern, Stephen. *The Culture of Time and Space, 1880–1918*. Cambridge: Harvard University Press, 1983.
Kern, Stephen. "Time and Art in Twentieth-Century Culture." In *Tempus Fugit: Time Flies*, Jan Schall, ed. Kansas City, MO: Nelson-Atkins Museum of Art, 2000. 22–34.
Newhall, Beaumont, ed. *Photography: Essays and Images*. London: Secker and Warburg, 1980. 59–60.
Olin, Margaret. "Touching Photographs: Roland Barthes's 'Mistaken' Identification." *Representations*, Vol. 80 (Fall 2002).
Orvell, Miles. "Reproduction and 'The Real Thing': The Anxiety of Realism in the Age of Photography." In *The Technological Imagination: Theories and Fictions*, Teresa de Lauretis, Andreas Huyssen, and Kathleen Woodward, eds. Madison: Coda Press Inc., 1980. 49–64.
Orvell, Miles. *The Real Thing: Imitation and Authenticity in American Culture, 1880–1940*. Chapel Hill: University of North Carolina Press, 1989.
Schall, Jan, ed. *Tempus Fugit: Time Flies*, Kansas City, MO: Nelson-Atkins Museum of Art, 2000.
Scott, Clive. *Street Photography: From Atget to Cartier-Bresson*. London and New York: I. B. Tauris, 2007.

Index

For Product Safety Concerns and Information please contact our EU representative GPSR@taylorandfrancis.com
Taylor & Francis Verlag GmbH, Kaufingerstraße 24, 80331 München, Germany

www.ingramcontent.com/pod-product-compliance
Lightning Source LLC
LaVergne TN
LVHW082333200926
848481LV00012B/3328
9781032338613